THE SCIENCE OF CAN AND CAN'T

Chiara Marletto is a research fellow at Wolfson College and the University of Oxford Department of Physics. Her main research focus is theoretical physics, but she also enjoys dabbling in theoretical biology, epistemology, and Italian literature. This is her first book.

* * *

Praise for *The Science of Can and Can't*

"[A] revolutionary recasting of physics . . . Marletto's contributions to 'constructor theory' reconcile what we think of as physical laws with the open-ended possibilities thrown up by biology and information theory. It is a paradigm that, for all its rigor, reenchants the world and enriches our place in it."   —*New Scientist*

"Marletto has a clear, sharp, and imaginative style of explaining science. . . . [*The Science of Can and Can't*] will open the doors to a dazzling, deep set of new concepts and ideas that will change and affect deeply the way you look at the world. Let her story unfold. It will be an open-ended exploration of the endless possibilities that the laws of physics allow for."
              —David Deutsch, author of *The Beginning of Infinity*

"I enjoyed this book very much, not least because of the freshness of its approach to a subject that can easily become hard for the nonscientific mind to grasp. The theory of 'can and can't' is an intriguing way of describing problems that are not only scientific (it describes very well what a storyteller does, for instance), and

Marletto's account of some things I thought I more or less understood (the nature of digital information, for one) illuminated them from an angle that showed them more clearly than I'd seen them before."
—Philip Pullman

"Chiara Marletto is trying to build a master theory—a set of ideas so fundamental that all other theories would spring from it. Her first step: Invoke the impossible."
—*Quanta Magazine*

"Hugely ambitious, Chiara Marletto is the herald for a revolutionary new direction for physics. This book is essential reading for anyone concerned with the future of physics."
—Lee Smolin, author of *The Life of the Cosmos*

"[A] cerebral yet intellectually satisfying journey with a simple description of the two kinds of counterfactuals in physics . . . Marletto's style resembles a frank conversation with the reader. Sophisticated concepts in physics, like information and knowledge, are explained using clear analogies to everyday life."
—*Booklist*

"[A] lyrical yet complex debut . . . References to Greek mythology, Shakespeare, chess, and LEGO add life to her survey. . . . Marletto's love of physics shines through. . . . Those with an interest in physics will appreciate her passion and her provocative approach."
—*Publishers Weekly*

"Chiara Marletto writes well about deep issues. I particularly like her suggestion that the current impasse in attempts to unite gravity and quantum mechanics might be broken if we concentrate on what things the two theories tell us can and can't be done."
—Julian Barbour, author of *The Janus Point*

The
Science
of Can and
Can't

**A Physicist's Journey through
the Land of Counterfactuals**

Chiara Marletto

PENGUIN BOOKS

PENGUIN BOOKS
An imprint of Penguin Random House LLC
penguinrandomhouse.com

First published in the United States of America by Viking,
an imprint of Penguin Random House LLC, 2021
Published in Penguin Books 2022

Illustrations drawn by Vlatko Vedral

ISBN 9780525521945 (paperback)

THE LIBRARY OF CONGRESS HAS CATALOGED THE HARDCOVER EDITION AS FOLLOWS:
Names: Marletto, Chiara, author.
Title: The science of can and can't : a physicist's journey
through the land of counterfactuals / Chiara Marletto.
Description: New York : Viking, [2021] |
Includes bibliographical references and index.
Identifiers: LCCN 2020037751 (print) | LCCN 2020037752 (ebook) |
ISBN 9780525521921 (hardcover) | ISBN 9780525521938 (ebook)
Subjects: LCSH: Science—Philosophy. | Natural selection. | Counterfactuals (Logic)
Classification: LCC QH375 .M37 2021 (print) |
LCC QH375 (ebook) | DDC 501—dc23
LC record available at https://lccn.loc.gov/2020037751
LC ebook record available at https://lccn.loc.gov/2020037752

Printed in the United States of America
1st Printing

Designed by Alexis Farabaugh

To Piera and Giuseppe

Contents

Foreword

This is an intensely rational, transformative, and delightfully humane book about the power of taking *counterfactual* explanations of the world seriously. Those are explanations about what physical events could or could not be *made to happen*.

This is a major departure from the traditional conception of physics and science more generally, which takes for granted that scientific theories can only be about what *must* happen in the universe (or what is likely to), given what *has* happened, and which rejects such intangibles as causation, free will, and choice as being mere psychological props, or even mystical. And it even classifies such essential laboratory concepts as temperature, information, and computation as being incompatible with any exact description of nature, and convenient only at the level of human sensory experience.

But none of that is true. Those are no more than arbitrary limitations on our ability to understand the world, adopted only by custom and habit. Fortunately, they are widely flouted both in everyday life and in theoretical science—albeit often guiltily and apologetically. If something is incompatible with the traditional conception, that does not make it incompatible with exact scientific description. It's just that in those cases, exact descriptions require a departure from the traditional conception—it requires counterfactuals. Something can hold *information* only if its state *could have been otherwise*: A computer memory is useless if all the changes in its contents over time are predetermined in the factory. The user could store nothing in it. And the same holds if you replace 'factory' with the Big Bang.

In this book, you will read why escaping from the traditional conception, and incorporating counterfactuals on an equal footing with factual statements into fundamental physics, is so promising—how it sheds a scientific light on much more of the world, informing a deeper conception of it and ourselves, and how it could facilitate further discoveries.

But there's more to it than that. Not only can counterfactuals provide new explanations; they are the foundation of a new *mode* of explanation. In the nineteenth and early twentieth centuries, not only were many new scientific explanations discovered, but new modes of explanation and understanding were being invented, too—such as evolution by natural selection, force fields, curved spacetime, quantum superposition, and the universality of computation. In the past few decades, in contrast, there have been none. Although new types of elementary particles have been discovered— and, for instance, the discovery of the Higgs particle was undoubtedly a triumph of both experiment and theoretical explanation—no new mode of explanation about physical phenomena has been discovered. In the first half of the twentieth century, however, the very idea of a particle as previously conceived had been swept away and replaced by a new, more deeply explanatory one.

With far fewer physicists, there were triumphs unmatched by anything in recent decades. Though the overall rate of scientific discovery has greatly increased by almost any measure, the discovery of *fundamentally* new truths about nature has, paradoxically, become less frequent. In fundamental physics in particular, there has been less and less exploration of transformative ideas—and new modes of explanation are not even being attempted.

This has happened for all sorts of more or less accidental reasons. But the net effect is a cautious and risk-averse culture in science. a preference for incremental over fundamental innovation, and for research with modest but foreseeable outcomes. In regard to fundamental progress itself, pessimism and fatalism have become the norm.

I don't agree with those who say that physics has already discovered all the "low-hanging fruit", and that all that remains is to harvest the rest, stolidly and mechanically. Nor with those who say that we apes are incapable of comprehending anything more fundamental than our current best theories, such as quantum theory and general relativity. On the contrary, in reality there has never been a time when there have been more blatant contradictions, gaps, and unresolved vagueness in our deepest understanding of nature, or more exciting prospects to explore them. Sometimes this will require us to adopt radically different modes of explanation.

The Science of Can and Can't sets out in nontechnical terms a new, counterfactual mode of explanation based on scientific and philosophical ideas that the author, Chiara Marletto, and I have pioneered. They provide new tools and new principles to address a number of notorious problems in physics and beyond. With a light but sure touch, Chiara Marletto argues for an emerging new theory, including a corpus of new and updated laws of nature— principles that will inform not just the next generation of atom-scale heat engines and nanorobots, but also artificial intelligence. This book goes through these topics with great enthusiasm and precision, punctuating the nonfiction in the chapters with short fictional stories that, in a manner reminiscent of Douglas Hofstadter's *Gödel, Escher, Bach*, elaborate the ideas of the book, to give the reader space to reflect.

In Chiara's land of counterfactuals, you will find new concepts (such as laws about information and knowledge) and old ones (such as work and heat) expressed in a radically different way. The Science of Can and Can't *can* enrich your understanding of the world, and of understanding itself.

David Deutsch

A Note on How to Read This Book

The details I remember most vividly from my childhood books are their illustrations: a colour plate depicting a monstrous whale in a memorable edition of *Pinocchio*; a black-and-white drawing of Captain Flint walking down a dark, narrow alley in *Treasure Island*; Quentin Blake's delicate illustrations for *Matilda*; Gustave Doré's terrifying beasts in Dante's *Inferno*; and so on. As I was writing this book, I thought I'd add a little something to each chapter that could work like illustrations do in books. Something as light and intriguing as a good conjuring trick; something that would capture the basic elements of the chapter and make them more memorable—turning them into a story. These stories are not essential to understand the scientific content of the book—you can skip them if you are in a hurry, and come back to them at some later point. They are intended as places where you can rest for a while, should you feel like having a pause from the scientific discourse that takes place within the chapters. I hope they will provide a fun addition to your exploration.

Prelude

f you could soar high in the sky, as red kites often do in search of prey, and look down at the domain of all things known and yet to be known, you would see something very curious: a vast class of things that science has so far almost entirely neglected. These things are central to our understanding of physical reality, both at the everyday level and at the level of the most fundamental phenomena in physics yet they have traditionally been regarded as impossible to incorporate into fundamental scientific explanations. They are facts not about what is the 'actual'—but about what *could or could not be*. In order to distinguish them from the actual, they are called *counterfactuals*.

Suppose that some future space mission visited a remote planet in another solar system, and that they left a stainless-steel box there, containing among other things the critical edition of, say, William Blake's poems. That the poetry book is subsequently sitting somewhere on that planet is a *factual* property of it. That the words in it *could be read* is a counterfactual property, which is true regardless of whether those words will ever be read by anyone. The box may be never found; and yet that those words could be read would still be true—and laden with significance. It would signify, for instance, that a civilisation visited the planet, and much about its degree of sophistication.

To further grasp the importance of counterfactual properties, and their difference from actual properties, imagine a computer programmed to produce on its display a string of zeroes. That is a factual property of the computer, to do with its actual state—with what is. The fact that it *could be reprogrammed* to output other strings is a counterfactual property of the computer. The computer may never be so programmed; but the fact that it *could* is an essential fact about it, without which it would not qualify as a computer.

The counterfactuals that matter to science and physics, and that have so far been neglected, are facts about what *could or could not be made to happen* to physical systems; about what is *possible* or *impossible*. They are fundamental because they express essential features of the laws of physics—the rules that govern every system in the universe. For instance, a counterfactual property imposed by the laws of physics is that it is *impossible* to build a perpetual motion machine. A perpetual motion machine is not simply an object that moves forever once set into motion: it must also generate some useful sort of motion. If this device could exist, it would produce energy out of no energy. It could be harnessed to make your car run forever without using fuel of any sort. Any sequence of transformations turning something without energy into something with energy, without depleting any energy supply, is impossible in our universe: it could not be made to happen, because of a fundamental law that physicists call the *principle of conservation of energy*.

Another significant counterfactual property of physical systems, central to thermodynamics, is that a steam engine is *possible*. A steam engine is a device that transforms energy of one sort into energy of a different sort, and it can perform useful tasks, such as moving a piston, without ever violating that principle of conservation of energy. Actual steam engines (those that have been built so far) are factual properties of our universe. The *possibility* of building a steam engine, which existed long before the first one was actually built, is a counterfactual.

So the fundamental types of counterfactuals that occur in physics are of two kinds: one is the *impossibility* of performing a transformation (e.g., building a perpetual motion machine); the other is the *possibility* of performing a transformation (e.g., building a steam engine). Both are cardinal properties of the laws of physics; and, among other things, they have crucial implications for our endeavours: no matter how hard we try, or how ingeniously we think, we cannot bring about transformations that the laws of physics declare to be impossible—for example, creating a perpetual motion machine. However, by thinking hard enough, we can come up with more and better ways of performing a possible transformation—for instance, that of constructing a steam engine—which can then improve over time.

In the prevailing scientific worldview, counterfactual properties of physical systems are unfairly regarded as second-class citizens, or even excluded altogether. Why? It is because of a deep misconception, which, paradoxically, originated within my own field, theoretical physics. The misconception is that once you have specified everything that exists in the physical world and what happens to it—all the actual stuff—then you have explained everything that can be explained. Does that sound indisputable? It may well. For it is easy to get drawn into this way of thinking without ever realising that one has swallowed a number of substantive assumptions that are unwarranted. For you can't explain what a computer is solely by specifying the computation it is actually performing at a given time; you need to explain what the *possible* computations it *could* perform are, if it were programmed in *possible* ways. More generally, you can't explain the presence of a lifeboat aboard a pirate ship only in terms of an actual shipwreck. Everyone knows that the lifeboat is there because of a shipwreck that *could happen* (a counterfactual explanation). And that would still be the reason even if the ship never did sink!

Despite regarding counterfactuals as not fundamental, science has been making rapid, relentless progress, for example, by developing new powerful theories of fundamental physics, such as quantum theory and Einstein's general relativity; and novel explanations in biology—with genetics and molecular biology—and in neuroscience. But in certain areas, it is no longer the case. The assumption that all fundamental explanations in science must be expressed only in terms of what happens, with little or no reference to counterfactuals, is now getting in the way of progress. For counterfactuals are essential to a number of things that are currently explained only vaguely in science, or not explained at all. Counterfactuals are central to an exact, unified theory of heat, work, and information (both classical and quantum); to explain matters such as the appearance of design in living things; and to a scientific explanation of knowledge. As I shall explain in this book, some of these things, such as information, heat, and work, already have some explanation in physics, but it is insufficient: it is only approximate, unlike more fundamental theories of physics, such as quantum theory and general relativity. Some others, such as knowledge creation, do not even have a fully fledged explanation yet. All these entities must be understood, without approximations, for science to make new progress in all sorts of fields—from fundamental physics to biology, computer science, and even artificial intelligence. Counterfactuals are essential to understand them all.

This book is an exploratory journey through the land of counterfactuals. It charts the territory beyond the boundary that has been currently set by the traditional conception of physics. Reading this book will feel a bit like going on an expedition in forbidden seas—like that of Darwin on the *Beagle*. You are going to explore alien lands with diverse beasts and creatures, taking note of what they are, and how they behave. At the end, you will have learnt something new about how to approach counterfactuals and how they are key to address long-unsolved problems. Most important, you will see that a barrier has been erected that prevents us from understanding those entities; that each counterfactual property is at the heart of fields where physics and science more broadly are currently unable to make actual progress; and that it is vital to trespass the boundary in order to incorporate them into physics and science.

To clarify how to do that, I shall describe a few key open problems in physics that can be resolved fully by using counterfactuals. I shall start with the most fundamental phenomena—classical and quantum information—and then proceed to the theories of life and knowledge, and finally consider thermodynamics. All these phenomena have something in common: they are currently at best only approximately expressed in physics, but a counterfactual-based science can provide a unified explanation of them all, while also revealing unexpected connections between them.

I shall also explain how new scientific theories about counterfactuals can be formulated; I shall outline a whole new way of describing the workings of the universe, which constitutes what I call the Science of Can and Can't. This counterfactual-based approach to science can dramatically overhaul our current way of looking at the world, making it sharper and more powerful. It is a mind-blowing step with the potential to unlock centuries-old secrets.

1.

Such Stuff As Dreams Are Made On

Where I explain how to look at the laws of physics in a far broader way, including **counterfactuals** (statements about what transformations are possible or impossible); and you become acquainted with **knowledge**—defined objectively, via counterfactuals, as information that is capable of perpetuating its own existence.

Most things in our universe are impermanent. Rocks are inexorably abraded away; the pages of books tear and turn yellow; living things—from bacteria, to elephants, to humans—age and die. Notable exceptions are the elementary constituents of matter—such as electrons, quarks, and other fundamental particles. While the systems they constitute do change, those elementary constituents stay unchanged.

Entirely responsible for both the permanence and the impermanence are the laws of physics. They put formidable constraints on everything in our universe: on all that has occurred so far and all that will occur in the future. The laws of physics decree how planets move in their orbits; they govern the expansion of the universe, the electric currents in our brains and in our computers; they also control the inner workings of a bacterium or a virus; the clouds in the sky; the waves in the ocean; the fluid, molten rock in the glowing interior of our planet. Their dominion extends even beyond what actually happens in the universe to encompass what *can*, and *cannot*, be made to happen. Whatever the laws of physics forbid cannot be brought about—no matter how hard one tries to realise it. No machine can be built that would cause a particle to go faster than the speed of light, for instance. Nor, as I have mentioned, could one build a perpetual motion machine, creating energy out of no energy—because the laws of physics say that the total energy of the universe is conserved.

The laws of physics are the primary explanation for that natural tendency for things to be impermanent. The reason for impermanence is that the laws of physics are not especially suited for preserving things other than *elementary components*. They apply to the primitive constituents of matter, without being specially crafted, or designed, to preserve certain special aggregates of them. Electrons and protons attract each other—it is a fundamental interaction; this simple fact is the foundation of the complex chemistry of our body, but no trace of that complexity is to be found in the laws of physics. Laws of physics, such as those of our universe, that are *not* specially *designed*, or tailored, to preserve any-thing in particular, aside from that elementary stuff, I shall call *no-design laws*. Under no-design laws, complex aggregates of atoms, such as rocks, are constantly modified by their interactions with their surroundings, causing continuous small changes in their structure.

From the point of view of preserving the structure, most of these interactions introduce errors, in the form of small glitches, causing any complex structure to be corrupted over time. Unless something intervenes to prevent and correct those errors, the structure will eventually fade away or collapse. The more complex and different from elementary stuff a system is, the harder it is to counteract errors and keep it in existence. Think of the ancient practice of preserving manuscripts by hand-copying them. The longer and more complex the manuscript, the higher the chance that some error may be performed while copying, and the harder it is for the scribe to counteract errors—for instance, by double-checking each word after having written it.

Given that the laws of physics are no-design, the capacity of a system to maintain itself in existence (in an otherwise changing environment) is a rare, noteworthy property in our universe. Because it is so important, I shall give it a name: *resilience*.

That resilience is hard to come by has long been considered a cruel fact of nature, about which many poets and writers have expressed their resigned disappointment. Here is a magisterial example from a speech by Prospero in Shakespeare's *Tempest*:

Our revels now are ended. These our actors,
As I foretold you, were all spirits, and
Are melted into air, into thin air:
And like the baseless fabric of this vision,
The cloud-capp'd tow'rs, the gorgeous palaces,
The solemn temples, the great globe itself,
Yea, all which it inherit, shall dissolve,
And, like this insubstantial pageant faded,
Leave not a rack behind. We are such stuff
As dreams are made on; and our little life
Is rounded with a sleep.

Now, those lines have such a delightful form and rhythm that, on first reading, something important may go unnoticed. They present only a narrow, one-sided view of reality, which neglects fundamental facts about it. If we take these other facts into consideration, we see that Prospero's pessimistic tone and conclusion are misplaced. But those facts are not immediately evident. In order to see them, we need to contemplate something more than what spontaneously happens in our universe (such as impermanence, occasional resilience, planets, and the cloud-capped towers of our cities). We shall have to consider what can, and cannot, be made to happen: the *counterfactuals*—which, too, as I said, are ultimately decided by the laws of physics.

The most important element that Prospero's speech neglects is that even under no-design laws, resilience *can* be achieved. There is no guarantee that it shall be achieved, since the laws are not designed for that; but it *can* be achieved because the laws of physics do not forbid that. An immediate way to see this is to look around a bit more carefully than was possible in Shakespeare's time. There are indeed entities that are resilient to some degree; even more importantly, some are more resilient than others. Some of them very much more. These are not, contrary to what proverbs and conventional wisdom might suggest, rocks and stones, but living entities.

Living things in general stand out as having a much greater aptitude to resilience than things like rocks. An animal that is injured can often repair itself, whereas a rock cannot; an individual animal will ultimately die, but its species may survive for much longer than a rock can.

Consider bacteria, for example. They have remained almost unchanged on Earth for more than three billion years (while also evolving!). More precisely, what has remained almost unchanged are some of the particular sequences of instructions that code for how to generate a bacterium out of elementary components, which are present in every bacterial cell: a *recipe*. That recipe is embodied in a DNA molecule, which is the core part of any cell. It is a string of chemicals, of four different kinds. The string works exactly like a long sequence of words composed of an alphabet of four letters: each word corresponds roughly to an instruction in the recipe. Groups of these elementary instructions are called 'genes' by biologists.

It is the particular structure, or pattern, of bacterial DNA that has remained almost the same over such a long time. In contrast, during the same period, the arrangement and structure of rocks on Earth have profoundly changed; entire continents have been rearranged by inner movements taking place underneath the Earth's crust. Suppose some aliens had landed on Earth early in prehistory, collected DNA from certain organisms (say, blue-green algae), and had also taken a picture of our planet from space; and that they were to come back now to do the same. In the pictures of the planet, everything would have changed. The very arrangement of continents and oceans would be utterly different. But the structure of the DNA from those organisms would be *almost unchanged*. So, after all, certain things in our universe, like recipes encoded in DNA, *can* achieve a rather remarkable degree of resilience.

The other element that Prospero's speech disregards is that living entities can operate on the environment, transform it, and (crucially) preserve the ability to do so again and again, thus leaving behind much more than 'a rack'. The Earth still bears the signs of bacterial activity from a billion years ago (for instance, in the form of fossil carbon). Plants have caused a dramatic change in the composition of the atmosphere by releasing gaseous oxygen as a side effect of converting the sun's light into chemical energy via photosynthesis. Humans, too, are capable of transforming the environment in a wide set of conditions. Contrary to Prospero's view, palaces, temples, and cloud-capped towers can achieve resilience—

because they are products of civilisation. Humans can restore them by following a blueprint—or rather, again, a recipe—of how they were initially built, guaranteeing that they will endure much longer than their constituent materials. In principle, a 3-D printer provided with such a recipe could reconstruct from scratch any ancient palace that happened to be completely destroyed.

The human life span may be still constrained, but technology has already extended it well beyond that of our ancestors. By changing the naturally occurring environment, human civilisation is tentatively improving and growing. We now have the knowledge to produce warm (or cooled) houses, powerful medications, efficient transport on Earth and even into space, and tools to save ourselves labour, to lengthen our lives and make them more enjoyable. We have majestic works of art and literature, music, and science. Those very words in Prospero's speech are an example of our literary heritage, and they have therefore survived—together with countless other wondrous outputs of human intellectual activity. So, rather than fading away, this pageant we have set up, which sustains us, has been under way for centuries. The rest of life's show on Earth has endured even longer, for billions of years.

Of course, the resilience of our civilisation is constantly threatened by severe problems, which crop up as we try to move forward. Some of them, such as global warming and fast-spreading pandemics, are in fact a by-product of the very progress I have described. These problems present considerable challenges and could easily wipe out several aspects of the progress we have made.

But the point I would like to focus on here is this: It is *possible* to take steps to solve those issues, no matter how serious they appear; and the laws of physics do not forbid still greater improvement. They do not guarantee improvement or resolution, but nor do they forbid it: resilience and further progress, by addressing problems such as the climate crisis, are both possible. The laws of physics, expressed as counterfactuals, offer a chance for improvement. By contemplating what is *possible* in the universe, in addition to what happens, we have a much more complete picture of the physical world. Prospero's gloomy conclusion is therefore partial and profoundly misguided. It was nothing more than an unreal nightmare.

These reflections suggest that the recipe in certain DNA patterns is much more resilient than stone; and that the elements of our civilisation for which there exists an analogous recipe, such as medicine, science, and literature, can be more resilient still. So, under no-design laws, a high degree of resilience seems to require there to be recipes of a particular kind. What kind? And what are such recipes made of, exactly?

The answer has to be constructed gradually and requires a digression about recipes. First, let's understand how recipes can be created under no-design laws of physics. After all, as I said, the only things that these laws preserve 'for free' are certain elementary particles and chemicals; one therefore has to understand how those recipes can have come about at all, out of elementary things that know nothing about recipes of such complexity.

I shall start with the recipes coded in the pattern of living cells' DNA. It is now well understood how those have come about. Darwin's theory of evolution explains how living entities and their stupendous biological adaptations—such as the snout of a dog, the fins of a dolphin, or the wings of a bee—have come about in the absence of a designer, under no-design physical laws. Now, each biological adaptation of a given animal is coded for somewhere in the recipe embodied in the DNA of that animal. What Darwin's theory tells us is how the recipes coding for complex biological adaptations can have come about without being explicitly designed. This will be key to understanding what the recipes are made of.

As is often the case with deep theories, grasping exactly what problem Darwin's theory addresses requires some excavation. The problem was stated with great clarity by the theologian William Paley a few decades before Darwin's breakthrough. Living things are so perfectly orchestrated that they seem to have been the output of an actual design process—such as that which produces a car in a factory—directed towards a purpose. They have the 'appearance of design', just like cars or smartphones or a watch. If you are walking along the beach and you suddenly see a watch on the ground, you may be guessing that some designer must have assembled it. But at the dawn of our planet's history there was no designer, factory, or intentional design process that could create living things: only elementary components of matter, served in the form of an amorphous bubbling soup, and nothing more. So how can living entities, and the resilient recipes coding for the biological adaptations in their structure, have come about in the absence of a designer?

What Darwin discovered, and what Paley could not quite see, is that there is no need for any intentional design process: biological adaptations in animals can be created out of elementary components of matter, such as simple chemicals, via a nonpurposeful process called *natural selection*. That process needs only enough time and elementary resources, such as simple chemicals and so on. It is an undirected mechanism, and yet it can produce purposeful complexity, starting from scratch under laws of physics that are simple and no-design themselves.

There are two key concepts in Darwin's powerful explanation (as it is understood today). One is that of a *replicator*—whose key role in evolution has been exposed with superb clarity by Richard Dawkins in the celebrated book *The Selfish Gene*. Think of the bacterial example again. Each instruction in the recipe to build a bacterium is embodied in a particular pattern of a portion of bacterial DNA; that portion is called a 'gene'. Now, genes have a special property. Every time the bacterium cell self-reproduces and creates a new instance of itself, each gene's pattern gets replicated, or copied, accurately; then the rest of the new cell is constructed by executing the recipe in the DNA. Since they are capable of being replicated, those patterns are called 'replicators'. Incidentally, their replication is a step-wise, 'letter-by-letter' process, similar to that used by monastic scribes to copy the content of ancient manuscripts; and it can be error-corrected via a similar method, which in bacteria is implemented by the cell once the replication has happened. In this way, the structure of bacterial DNA has survived for long: by being copied from generation to generation and potentially preserved for a much longer time than the bacterium's life span thanks to error-correction enacted by the cell. It is interesting that what's passed on from generation to generation, via replication, is the particular pattern that codes for a gene, or an elementary instruction: every time it is copied it changes its physical support, while retaining all its properties as a pattern. It is the same as what happens to the sequence of words copied by the scribes: the ink and bits of parchment embodying those words change, but the copied words are, if no typos occur, the same as those in the source manuscript. Patterns with this particular counterfactual property, that of being *copiable* from one physical support to another while retaining all their defining properties, are a special case of 'information'—of which I shall give a precise explanation (based on counterfactuals) in chapter 3.

So, the resilient recipes we see in animals around us must be constituted by some kind of information. To understand what kind, you must consider the second powerful concept of Darwin's theory: variation and selection.

While the copying process occurs from generation to generation, since the physical laws are no-design, errors can happen as a result of the interaction with the environment: these result in nonpurposeful changes ('variations') in the replicators. When errors happen at the right pace, not too often, and not too rarely, they produce novel variants of the genes in the newly formed bacterial cell, coding for a different trait—they produce new recipes. Sometimes, this means that individuals with that variant are able to cope better with the environment and become more successful—thus granting the perpetuation of that variant gene, and the recipe it codes for, to the detriment of the others. Less successful variants eventually go extinct as a result of the competition with the more successful variants in that particular environment. This phenomenon is 'natural selection': the blind process that can bring about something as graceful as a winged butterfly, and as elegant as an inky black panther, without having any clue as to what they should be like, just because replicating molecules replicate.

Natural selection gives us the key to explaining what makes the information in the surviving recipes worthy of attention. Since natural selection is blind, only few, particular changes are valuable and generate replicators that are capable of keeping themselves in existence: most of them are not, and lead to extinction. For example, in a forest where the trees have dark-coloured barks, only certain changes to the genes coding for the pigment of a moth's wing would be advantageous: for example, those that make the pigment closer to the bark's colour, so as to make the moth carrier of that trait less visible to predators. What distinguishes helpful changes in the recipe from unhelpful ones? It is a particular kind of information: information that is *capable* of keeping itself instantiated in physical systems. It is resilient information.

I shall call this resilient information, which is the ingredient in successful recipes, 'knowledge' (and I shall talk about it extensively in chapter 5): for adaptations, it is knowledge of some features of the environment, such as that trees have dark-coloured bark. Knowledge in this sense does not have to be known by anyone: the moth does not know its wings are black. 'Knowledge' merely denotes a particular kind of information, which has the capacity to perpetuate itself and stay embodied in physical systems—in this case by encoding some facts about the environment. Natural selection is a process that, by nonpurposefully selecting for biological adaptations, can accidentally create knowledge. It is a nondirected, blind process, which with enough time and generic resources can bring about things that look as if they had been designed. But that is an illusion: no designer is needed.

The other kind of recipes I mentioned is those that maintain our civilisation in existence—by coding for how to build things like palaces, factories, cars, and robots.

Such recipes contain knowledge, too: they consist of information that can perpetuate itself, embodied in physical supports such as our brains, bits of papers, books, documentaries, historical records, scientific papers, conference proceedings, the internet, and so on. However, this kind of knowledge is brought about via a process different from natural selection: it is produced by thinking, and it can reach further than knowledge that emerges directly by natural selection.

It is primarily via this kind of knowledge that humans have been able to construct a civilisation that is tentatively improving and growing, despite also often making bad mistakes. Such knowledge consists of thoughts. It is made of the same stuff as dreams are made on. Yet rather than fading away like fog in the morning sun, as Prospero suggests, knowledge is the key to resilience. The knowledge in his speech survives to this day. In fact, knowledge is the most resilient stuff that can exist in our universe.

Given that knowledge has such an essential role in the survival of complex entities, it is essential to understand the process by which new knowledge is created from scratch in our mind. Fortunately, this process was elucidated by the philosopher Karl Popper in the mid-twentieth century. He argued that knowledge creation always starts with a problem, which we can think of as a clash between different ideas someone has about reality. Incidentally, this suggests a rather positive, uplifting interpretation of conflicting states of mind where contrary impulses clash and fight. These conflicts are all examples of problems: but luckily problems can lead to new discoveries. For example, when writing a story, the clash in the author's mind might be between the desire to use elegant, lyrical language and the necessity of keeping the attention of the reader alive with a gripping plot. The author has to find a way of meeting both these criteria, which may clash in certain situations: a long passage describing an idyllic landscape might give a perfect chance to meet the former criterion but might result in the reader dropping the book and switching on the TV, because it slows down the pace of storytelling. To address problems such as this, one has to create new knowledge.

First, one conjectures several tentative solutions—the analogue of variations in replicators in natural selection. These could take the form of actual drafts written down on paper, or just thoughts, or a combination of those. Those conjectures may well be full of errors and produce even worse results at first. So, one proceeds with the second phase: criticism. Criticism is the act of seeking and correcting errors in an attempt to improve on the solutions the analogue of natural selection. Sometimes this process may be completely opaque to us, so that we may have the impression that good ideas come out of the blue. But it does in fact take place. The author will usually discard most of the early versions of the story until some final product that meets both criteria comes about. This final product, if the process has worked, may have the hallmark of all masterpieces: it is hard to change further, while still meeting the criteria, because it has been obtained by tentatively removing flaws in previous versions, which met the criteria to a lesser extent. The masterpiece contains new knowledge: it shall be remembered; it shall be translated into different languages; it shall live on for centuries and survive for generations, inspiring readers of all ages, as long as civilisation survives. As Shakespeare's Sonnet 18 says of itself:

So long as men can breathe, or eyes can see,
So long lives this, and this gives life to thee.

That process is tentative: given that there is no design in the laws of physics, there is no guarantee that one shall make progress by conjecture and criticism. But one *can*. For the same reason, a solution that looks good for one problem may at a later stage be found inadequate. For example, in physics, Newton's theory of gravitation had been tremendously successful, for nearly three centuries, at explaining planetary motion and many other things, but nevertheless it was later found to be inadequate and was superseded by a better theory, Einstein's general relativity. There are no absolute sources of certain truth: any good solution to a problem may also contain some errors. This principle is based on *fallibilism*, a pillar of Popper's explanation of rational thinking. Fallibilism makes progress feasible because it allows for further criticism to occur in the future, even when at present we seem to be content with whatever solution we have found. It leaves space for creating ever-improving theories, stories, works of art, and music; it also tells us that errors are extremely interesting things to look for. Whenever we try to make progress, we should hope to find more of them, as fast as possible.

So far, I have explained that recipes are key to resilience; that they are made of a special kind of information—knowledge—which has the ability to keep itself in existence. I've also explained how the two known processes of creating knowledge work: by conjecture and criticism, in the mind; by variation and natural selection, in the wild.

An important point is that the laws of physics allow for knowledge creation, but they do not guarantee it. So knowledge creation may stop at any point. For example, natural selection can sometimes enter stagnation—which can result in events such as mass extinctions (like those that took place in prehistoric ice ages). The reason is primarily that natural selection, unlike conjecture and criticism, cannot perform jumps: each of the recipes that leads to a new resilient recipe must itself be resilient—i.e., it must code for a successful variant of a trait of the particular animal in question that permits the animal's survival for long enough to allow replication of that recipe, via reproduction. But there may be viable, resilient recipes coding for useful traits that can never be realised because they would require a sequence of nonresilient recipes to be realised first, which is impossible, as those recipes produce animals that cannot survive and cannot pass on their genes.

The thinking process, in contrast, can perform jumps. As we all know very well, the sequence of ideas leading to a good idea need not consist entirely of good, viable ideas. Nonetheless, knowledge creation in the mind, too, can enter stagnation and stop progressing. We must be wary of not entering such states both as individuals and as societies. Particularly detrimental to knowledge creation are the immutable limitations imposed by dogmas, as they restrain the ability to conjecture and criticise.

The stupendous enterprise of knowledge creation has been unfolding over the centuries, producing brilliant works of art, music, literature, and powerful scientific theories. Looking back, we are comforted by seeing the progress made since the early beginnings of our civilisation. Looking ahead, it would seem that there might be several different fields in which progress can occur, especially in solving urgent and dramatic problems that humanity is now facing—such as the issues related to climate change and the open challenges in medicine and macroeconomics.

But if we could zoom out and see the arena of knowledge creation from above, a rather different scenario would appear within the domain of fundamental science, and of physics specifically. A boundary has been generated that affects and constrains the way criticism and conjectures can occur—a boundary that is keeping out certain kinds of explanations from the allowed set. These are explanations that involve counterfactuals. The boundary grew up because of a phenomenon that has been going on for some time, silently, largely unnoticed—like water seeping into a ship whose hull has a hidden hole below the waterline. To see what it is, we must start where it all began. We must start with physics.

It is perhaps ironic that this boundary-generating phenomenon started in physics, because physics is one of the clearest examples of how thinking can produce knowledge and make rapid progress. At a glance, from what one is taught in elementary courses at school, physics may appear like a collection of tools to solve irrelevant problems, of the kind you get in weekly physics tests: What is the time of flight of an apple that falls from a tree from a certain height? How long will it take for a bathtub of such a volume to be filled with water if the water is flowing in at this rate? And so on. Compared with other disciplines, such as literature or philosophy, physics may not seem to be about deep things at all. Who cares, after all, about how an apple falls? Isn't that fantastically narrow in scope?

This first impression is very far from the truth. Physics is a dazzling firework display; it is profound, beautiful, and illuminating; a source of never-ending delight. Physics is about solving problems in our understanding of reality by formulating explanations that fill gaps in our previous understanding. The point of physics is not the particular calculation about the fall of an apple. It is the explanation behind it, which unifies all motions—that of the apple with that of a planet in the solar system, and beyond. The dazzling stuff consists of explanations: for they surprise us by revealing things that were previously unknown and very distant from our intuition, with the aim of solving a particular problem. As I said, problems always consist of a contrast or clash between ideas about the world. For example, in the past, people believed that the Earth was at the centre of the universe; but that notion clashed irremediably with observations, such as those about the apparent movements of the stars, of the other planets, and of the Moon. This led Copernicus and Galileo to conjecture that the sun, not the Earth, was at the centre of the solar system. The Copernican Revolution was an astonishing change of perspective, which allowed us to make formidable progress in understanding astronomy and celestial mechanics, and eventually led, via a series of further steps, to our current space exploration enterprises.

By solving problems of that kind, physicists have gradually uncovered entirely unsuspected worlds, telling us a deeper layer of the story of how things are. These layers are beyond the immediate reach of our senses, but our mind can visualise them in the light of explanations.

In existing physics, all explanations have some primitive elements, in terms of which the physical reality to be explained is expressed. The appearance of the dark sky at night is a perfect example of that. It can be explained in terms of unexpected underlying phenomena involving things like photons, the remarkable fact that the universe is expanding, and so on. None of those elements is apparent in the sky itself, but they are all part of the explanation for why it looks as it does, in terms of what is really out there. Explanations are accounts of what is seen in terms of mostly unseen elements.

There is no limitation, in principle, to how deep one can go in finding even more primitive elements. The primitive elements of an explanation can always be explained further in terms of other entities, and so on, going deeper and deeper. Deeper levels of explanation may look very different from the shallower ones. For instance, there was a time in physics when particles were thought to be the ultimate elements of reality: these are discrete lumps of matter interacting with each other via forces at a distance. That view was then overturned by the idea of fields. A field in physics is a thing that permeates everything there is in a continuous way; particles are now understood as excitations—ripples—of fields. But fields themselves could, in principle, be broken down further, into more primitive elements of explanation—opening up a novel, and even more fundamental, explanation of reality. This may be hard to imagine for us, but we must be prepared to imagine more fundamental entities than fields, given that physics is open to further, more fundamental explanations. The resulting picture of scientific knowledge is that there are different levels of explanation about reality. Each of these levels may sometimes be autonomous, in the sense that it does not need to refer to the others to make sense of its own internal rules for example, it is still fine to think of particles without referring to fields if you wish to describe certain simple mechanical interactions, such as the collision of two rigid spheres. None of those levels is exhaustive. All the levels are essential to understanding what is out there.

The usual output of knowledge creation in physics is a piece of knowledge that addresses a particular problem: for example, the explanation for why the sky appears dark at night; the explanation for why the sun appears to move in the sky from east to west every day, and so on. From time to time, such problem-solving leads to an entirely new physical theory—such as Newtonian mechanics, general relativity, or quantum theory. These rare events have momentous consequences, resulting in a radical change in the way we look at the world, which may take several decades to be assimilated. Often, a new physical theory's practical and theoretical implications can be worked out only after a long while. For example, nothing in Einstein's theory of general relativity even hinted at GPS (the Global Positioning System), which provides information about location and time to our phones and cars using a network of satellites orbiting Earth. Yet GPS relies directly on the phenomena described by general relativity. The possibility of GPS is a counterfactual allowed by general relativity.

That's why a new physical theory is much more than a solution to a particular problem. It is a conjectured explanation that attempts to approximate the actual laws of physics—the rules that constrain everything in our universe. If you asked a physicist to write down what we currently know about the laws of physics, they would probably start writing a bunch of equations—for instance, $E = mc^2$. But then they would think again. They would start adding words to explain what those various symbols mean: E is energy, m is mass, c is the speed of light, and so on. And they would explain in words what energy, mass, speed, and light are. All those words constitute the explanation that is the core content of the physical theory that those equations express. The two ingredients are indissoluble: without explanations, an equation is empty and has no meaning. Without formulae, the explanation is too vague to be applied. A physical theory, therefore, is not just the set of its formulae, such as $E = mc^2$, nor is it just the collection of its testable predictions; it is a conjectured explanation, which includes, for example, the informal descriptions of what E, m, and c are in that formula, and why they are related in that way. This will also apply to things that, unlike the speed of light, cannot be directly measured, such as the 'geometry of spacetime', which are nevertheless crucial to explain why that formula (which is then relevant to make predictions) is as it is.

In practice, physical theories about the universe that count as viable explanations must at least have certain traits that guarantee they are free of basic flaws. In the first place, they must be *exact*. By 'exact theory', I do not mean expressible precisely in mathematical terms, or anything like that. I mean a theory that does not include any limitation as to the accuracy of its statements; in short, one that does not include any approximation. Think of two recipes for a cake, one requiring that you put *approximately* 100 grams of sugar in a bowl, the other requiring that you put in exactly 100 grams of sugar. The first is an approximate recipe, in that 99 grams or 101 grams will probably do; the second is an exact one. Just as with recipes, approximations in physical theories are vague as to what they say about physical reality, and for that reason they are problematic; for example, in regard to those recipes, one could ask why approximately 100 grams of sugar, and not exactly 99? An example of an exact physical theory is Newtonian physics, which allows one to predict the exact place and time of an apple's landing on the ground, once we know when and where it comes off the tree, and its initial velocity. Newtonian physics is also an example of the most general kind of theory—in that it is *universal*. A universal theory is one that is not subject to any limitation about its domain of applicability: Newton's theory applies to apples on Earth, and on Mars, and in any other alien place in the universe. Again, physical theories that are not universal and apply only to some scales, or domains, are by themselves problematic—because one still has to explain why they hold only at that scale and not elsewhere.

So, by tentatively solving problems in our understanding of physical reality, physics ends up seeking universal and exact physical theories. These theories, as I shall explain in chapter 2 and chapter 7, must also be *testable* so that they can be checked against reality to find potential errors.

Because of fallibilism, it is important to note that here 'exact' does not at all mean 'true'. Any conjectured explanation which seems to be working may be found to be false at any time. As I said, this happened with Newton's theory of gravitation when it was superseded by quantum theory and general relativity. We can never know whether a physical theory that we have formulated is true; all we can say is that it has so far not been found to be false. This may seem a little unsettling, but it is an extremely interesting fact about how knowledge is created; and, as I said, it is central to the possibility of making progress, via criticism.

Here we get closer to the origin of the pernicious boundary to exclude counterfactuals. As one can imagine, there are different ways in which explanations can be formulated. How many? We do not know: infinitely many, presumably. The mode of explanation of Newton's theory has a distinctive feature. Its scope is confined to explaining what happens in the universe, given two primitive elements: one includes what in physics jargon are called 'laws of motion'—the rules that tell us how the motion of systems (what physicists call the 'dynamics') unfolds in space and time; the other are the specific initial conditions of the motion. For example, Newton's laws of motion can be applied to say what happens to an apple, given the initial conditions (the particular place where the apple started its motion and its initial velocity). The set of points that a system goes through as it moves is called a 'trajectory'. Hit a tennis ball with a racket against a wall: the trajectory is the imaginary line one can draw to describe where the ball goes after it leaves your racket. The laws of motion and the initial condition give us a way to predict that trajectory without actually having to observe any actual ball being hit. Given the initial position and velocity of the ball, one can predict precisely where its motion will bring it, just computing the trajectory from the laws of motion.

This mode of explaining things in terms of what happens has proved extremely successful and far-reaching: it allows for powerful predictions, which can then be tested with experiments, to enable conjecture and criticism. This mode continued to be successful even when Newton's theory was found to be inadequate, by, for instance, failing to describe the precession of the planet Mercury: it delivered theories like quantum theory and general relativity, which are our current best theories to explain physical reality. Both of these subtle theories are formulated as laws of motion. It's the very same mode of explanation that Newton adopted in formulating his laws.

Along with much progress, this mode of explanation has, perversely, generated the wretched boundaries that could stand in the way of future successes. An unspoken stipulation was made—what I shall call the *traditional conception of fundamental physics*—that all fundamental physical theories must be formulated in terms of predictions about what happens in the universe, given the initial (or more generally, supplementary) conditions and laws of motion. In this conception, physics is no longer an open-ended enterprise. It has been infinitely narrowed to the project of finding theories that can be expressed only in terms of what happens in the universe, given the laws of motion of all its constituents, and a particular initial condition. So the ultimate theory about physical reality would consist of a collection of the trajectories of all elementary particles in the universe, given where and when they started. We do not have such a theory yet. But it is traditionally regarded, hypothetically, as the ultimate explanation of everything important about the universe.

That traditional conception has created the barriers against counterfactual explanations; and its project, if taken literally, ap-

pears impossible. In the first place, it is not possible to explain literally everything in terms of initial conditions and laws of motion:
for example, even if we had a decent theory of what the initial
conditions of the universe are, it could not itself be explained in
terms of initial conditions; for a start, it would have to contemplate
what would happen if other initial conditions were chosen—a
counterfactual explanation! How to explain the choice of the initial conditions is indeed an open problem in fundamental physics.
There are also other, related open issues that require that counterfactuals be addressed, such as the problem of "fine-tuning the laws
of physics"—about why dynamical laws are as they are (for an excellent exposition of this problem, see Paul Davies's *The Goldilocks
Enigma*). The fine-tuning problem cannot be addressed by stating
only what happens; one has necessarily to look at what might have
happened if the laws had been different. And how can one do that
without counterfactuals? In addition, explaining what we see now
in the universe around us in terms of a story that starts with given
initial conditions is itself arbitrary: one could describe everything
that happens, including what we see now, given the final conditions of the universe, and then use the laws of motion backwards,
by 'retrodicting', instead of predicting, the current state of affairs.

The traditional conception is also perverse because it clashes
with the pillars of rational thinking which I mentioned early on:
that of being changeable and improvable via conjecture and criticism. Physics aims at solving problems; as a consequence, it seeks,
if possible, universal and exact testable laws, formulated in what-

ever mode of explanation happens to be appropriate. In contrast, the traditional conception forces theories to come only in one form, thus narrowing down the space available for thinking. It introduces a boundary, which impedes progress. It confines physics only to things that can be described exactly in terms of statements about what happens given the initial conditions and laws of motion; but not about other phenomena, which thus remain only imperfectly explained.

There is more. The traditional conception of physics has inspired an approach that has now spread to other parts of science, too, via an approach that has been called 'reductionism'—the idea that there is only one level of explanation that is both fundamental and admissible, and everything else can be reduced to that. Such a level of explanation is, presumably, that of elementary particles or fields, and their trajectories, given their initial conditions. But this take on physical reality is, again, too narrow. There are questions that this approach cannot answer—questions that are deep and important for understanding the full reality of a physical phenomenon. For instance, the question "Why is a given transistor (in a computer) 'on' at the end of a given computation?" has at least two answers. One is that it is 'on' because the electrons in the computer were set in such-and-such initial conditions; the other is that it is 'on' because the computer performed a computation to find the factors of, say, the number 15, and the 'on' transistor is part of the encoding of the output (3 and 5). A reductionist would discard the latter explanation as 'less fundamental' (because, after all, "factoring a number is nothing but electric currents in the computer"). Reductionism ultimately denies that the computational description is necessary, though some reductionists may accept that it is helpful as a manner of discourse. But this is, of course, nonsense. Both explanations are essential to understand what is happening; they refer to different, autonomous levels of explanation, which do not implicate one another. By ignoring one of them, one misses something crucial about reality. Reductionism impedes progress in physics and in science in general because it requires all explanations to conform to certain arbitrarily predefined criteria—for instance, that they refer exclusively to microscopic particles and their trajectories.* In the example I gave earlier about computations, the explanation in terms of microscopic particles and their initial conditions (i.e., electron currents in the computer) is not enough to capture the full picture of what is going on (i.e., factoring a given number). Yet reductionists insist in dismissing whatever does not fit into those criteria (from information and thermodynamics, to creativity and consciousness) as 'approximate', and thus outside the scope of science. The result is a narrow and limited view of science.

---

*Sometimes rejecting reductionism is mistaken for giving in to nonscientific or (even worse!) supernatural explanations. This is not the view I am advocating: I am advocating that scientific explanations which do not fit the traditional, reductionist mode of explanation are necessary to understand physical reality; and that they should not be dismissed as irrelevant because they violate reductionism's central tenet.

There are phenomena that cannot be fully expressed by the traditional conception. By this I mean that physical theories and explanations about those phenomena can take only approximate, non-exact forms when expressed using the traditional conception's approach. So, by restricting oneself to that approach, one cannot adequately explain them within science.

One important example of things the traditional conception cannot adequately capture includes thermodynamic entities such as those associated to particular kinds of energy transfers—in physics, they are called 'work' and 'heat'. The laws stating how work can be turned into heat, and vice versa, are central to things like heat engines, which made possible the industrial revolution. Yet thermodynamics is often regarded as only a useful approximation, not a fundamental physical theory; so heat and work are regarded as not worthy of further explanation, because an exact physical theory about them cannot be cast in terms of statements about what happens, given initial conditions and laws of motion. The traditional conception has thus given up on an exact understanding of work and heat and similar entities, and claims to be content with the existing, problematic, approximate theories. These theories, as you will see, are highly effective, but only in certain limited domains (e.g., to design heat engines such as those used in cars and locomotives); however, they appear to rely on various approximations which, when we consider these laws as fundamental, become inadequate. I shall explain these issues, and how to solve them with counterfactuals, in chapter 6.

More generally, any phenomenon or entity that happens not to fit in an exact physical theory expressed by stating what happens given dynamical laws and initial conditions is considered by the traditional conception as an accessory phenomenon. Philosophers of science call these entities 'emergent' because they emerge as meaningful entities only at a certain level of explanation, which the traditional conception regards as nonfundamental. Even though our current understanding of them is not exact and poses problems, the traditional conception urges us to press on, declaring those entities as not really of interest to fundamental physics. The problem with this take is that *all* levels of explanations are necessary to grasp a given situation (remember the example with computation and factoring): levels of explanation work together like layers in a cake. It is impossible to get the cake's full flavour by ignoring the top layers and just sticking to the base. In this book you will be able to grasp the flavour of the full cake, by being introduced to counterfactuals.

To understand what exactly physics and science are missing by relying solely on the traditional conceptions, let's take a look at a simple example of some property that cannot be adequately captured by the traditional conception. To this end, I can evoke an imaginary masterly storyteller. As well as being dedicated to his craft, he is generous and enjoys featuring in stories that other authors write. For the sake of argument, suppose he was the best storyteller ever and was fond of writing novels by hand—say, with a green ballpoint pen. As a prolific writer, he likes always to have some blank paper at hand for when inspiration strikes, and he keeps, in a particular secret drawer of his desk, an emergency supply of blank sheets—a thick pile of them. Now, it may be the case that over the course of his whole career that special stash of paper is never used. In this case, the sheets will stay blank: this fact is something the traditional conception of physics could, in principle, predict, given the initial conditions of the universe and the laws of motion. However, the most important property of that paper is not that it will stay blank: rather, it is that *something could be written* on it. This property is the most important because it explains why the paper is kept there in the drawer, and why it is blank. Yet the traditional conception of physics cannot express it. That property is about what *could be made to happen* to the white paper. It is a counterfactual property: as I said earlier, it is about what could be, rather than what is. The traditional conception of physics cannot possibly capture counterfactual properties, because it insists on expressing everything in terms of predictions about what happens in the universe given the initial conditions and the laws of motion only—in terms of trajectories of apples or electrons, forgetting the other levels of explanation. But these other levels of explanation are essential sometimes to grasp the whole of physical reality. Neglecting them misses out on several aspects that are significant for a full understanding of what's going on.

Consider, as another example, a simple story: one that could be told from generation to generation by oral tradition, without having to be written down. An ancient Greek myth will do. The story goes that Theseus, son of Aegeus, king of Athens, went to Crete to kill the Minotaur. Theseus made an agreement with his aged father that if he defeated the Minotaur, on their return his crew would raise white sails on the ship; if he perished, they would raise black sails. So off went Theseus, and he defeated the Minotaur. But on his way back, distracted by all sorts of things (including, possibly, the presence of his fiancée, Ariadne, on the ship!), he forgot to tell the crew about the sails. The crew left the black sails on, and Aegeus, who from the highest tower of Athens could see the ship approaching, thought his son was dead. So he threw himself into the sea and drowned. This tragic story is why the sea is now called the Aegean.

Now suppose we asked our master storyteller to tell that story with the constraint that he can formulate statements only about what happens—that is, he must report the full story without ever referring to counterfactual properties. In particular, he cannot refer to properties that have to do with what could or could not be done to physical systems.

This task turns out to be impossible: for the story to make sense, and to convey fully its meaning, two attributes of the ship are essential: one, that *it can be used to send a signal,* by assuming one of two states—white sail showing or black sail showing; the other, that the state of having black or white sails *can be copied* onto other physical systems—such as Aegeus's eyes and brain. The copiability property tells us that the flag contains information (just as in the case of replicators).

These two properties, just like the property of blank paper, are counterfactual. So, that myth could not be told, conveying its full meaning, under the constraint that one should refer only to what happens. Not even by the best storyteller ever!

Likewise, by restricting itself to statements about initial conditions and laws of motion, physics is missing something essential about physical reality. Counterfactual properties are not a tiny curiosity in the realm of properties in the physical world. They are central for understanding a vast number of key facts about physical reality, which have been so far considered emergent and not fundamental. The consensus has now given up on understanding them properly with exact and universal laws. I have mentioned several such entities already in this chapter. The properties central to our modern understanding of the phenomenon of life are all counterfactual: replicators *can* get themselves replicated; the appearance of design is about what a particular aggregate of atoms *can* do when provided with appropriate inputs; resilience is the *capability* of a system to preserve itself. Information, which is about the *possibility of copying* certain patterns, is another counterfactual concept that is central to our technology. Knowledge, too, which is crucial to understanding phenomena like artificial intelligence, is defined by a counterfactual property: it is information that *can* keep itself in existence. Thermodynamic work and heat, which are key to our understanding irreversibility and energy transformations in the universe, are about counterfactual properties, too, as I shall explain in chapter 6. Even something as exotic as quantum computation—which you will encounter in chapter 4—has, at its heart, a number of exact counterfactual properties of physical systems. Far from being mysterious and spooky, as is usually said of quantum phenomena, these properties are exactly expressible and understandable.

In this book you will see that it is possible to formulate exact and universal laws having counterfactuals as their primitive elements, expressed via a different mode of explanation, that focuses on what can, and cannot, be made to happen. In the coming chapters, we shall be looking at how the Science of Can and Can't operates, expressing exact laws about those entities that the traditional conception has struggled to incorporate since the very beginning. How to develop that science is another story. Whether that story will be told some time in the future, and how good and resilient a story it will be, depends on the dream-like stuff we shall be able to create. It is up to us.

Making Sense

R ita dropped the pen on the physics notebook and huffed loudly. She had reached the end of a long sequence of multiple-choice questions. It felt like her mind had just turned into jelly. For a moment, she thought it was amusing to imagine her skull filled by a colourful, fruity, semisolid mass, with bits of physics equations randomly trapped in it. Then frustration settled back in. Her eyes were roving all over the room, in search of a way out.

"Dad?" she finally called.

Her father was sitting on the terrace nearby in a deck chair, wholly immersed in his reading. Through the half-open window *of her room, Rita could see only the top of his head, covered in dark curly hair, above the backrest.*

"Dad?" she repeated, raising her voice.

No reply. Rita wasn't surprised: her father often lost himself in his reading, becoming almost oblivious of his surroundings. She grabbed her physics notebook and joined him on the terrace. She cleared her throat. Her father finally raised his head and gave her an inquisitive look. "Oh, Rita. I thought you were doing your homework."

"Yes, I was. I think I've got all the answers more or less correct . . ." Rita hesitated and looked doubtfully at her notebook. "Hmm. I think I don't get on with physics much. I mean, I can do the calculations all right, but what's the point?"

"Well, it's physics . . ." said her father idly, his eyes back on his book, scanning the page in search of the line he had just dropped. "Physics is important, you know", he went on. "I may be an engineer, but, given the choice, I would have been a physicist."

"That engineering is worse than physics does not mean physics is good!" Rita exclaimed. To be fair, she hadn't the faintest idea about engineering. But she had intuitively placed it in the set of things collectively labelled as 'lethally boring stuff'. Physics belonged there, too.

Her father sighed, put the book down on his lap, and prepared himself for a long chat. "All right. What's the problem with this homework, exactly?"

"The problem is physics, I guess", Rita said timidly. "You see—look at these questions. First, they asked me to apply Newton's second law to derive the equation of motion of a projectile. Fine, I did that. Then they asked me to do the same with four other different initial conditions. The same stuff repeated, like, five times—five! Then, they wanted me to apply the third law. You know, the action and reaction rule. Something to do with a cannon resting on a bridge, or whatever. This stuff seems pointless. And the whole of physics is like this. I see no use for this subject. It seems so narrow. You see, other subjects do have a point. English, philosophy, art, and maths—for each one of them, I can tell you exactly what they are about—and they can get really fun. Even history, that I don't like much, has a point. But physics? I am not quite sure." She paused, thinking.

Rita's father seized the moment to steer the discussion towards a different direction. "Look", he said, "the way you are learning physics is not the best, I must admit . . ."

"Well, in my teacher's opinion, no one teaches physics better than him!" observed Rita bitterly. "Though he does not radiate any enthusiasm when he talks about physics. It seems to me he cares about the subject even less than I do."

Her father was almost amused, but he tried to look serious. "The point is that physics is not really the stuff you are learning", he pointed out gently. "I have talked to people who do physics for real, and they aren't solving problems like those. Not at all."

Rita stared at the floor, prickling with frustration.

"*Yes, they solve equations and compute things, and they surely learnt how to do that properly, at one point in their life, in school. But what they are really doing, behind those equations, is trying to understand the universe: how it works, what are its most fundamental rules, and things like that.*"

Rita raised her head a little and said flatly: "*I thought that was what philosophers do.*"

"*Sure, philosophers do that too—in their own way; they think about arguments on ethics, aesthetics, knowledge, and whatnot. But physicists are especially interested in finding out the very foundations of the universe—the laws that drive everything else. Those laws contain amazing ideas, just like well-crafted philosophical arguments sometimes do.*"

"*Are you sure?*" Rita was sceptical—she had seen none of that in her physics classes. But she trusted her father. Above all, she wanted to believe that was true. It would have made her life as a student so much better.

"*Well, no one is sure of anything in these philosophical discussions! But yes, I believe it's like that*", confirmed her father.

"*Hmm. That's odd: why would physicists still be looking for new laws? I thought we already knew all the laws of physics. These equations I'm studying, like Newton's laws and stuff—aren't they enough?*"

"*Because the laws of physics as we know them right now may not ultimately work. They are not settled at all! Newton's laws and others that were initially proposed are now obsolete; at some point,* they were found to be wrong. Now we have even more general laws. You will study them in a while, maybe at the university, depending on what you will choose to do ..."

Rita made a perplexed face. She was still unconvinced. "I thought that physics was made of some given set of laws. Fixed rules, given once and for all, and doing physics is just learning them and using them. But you're saying it's not like that?"

"Yes. Physics is far from being settled. As I said, Newton's laws used to be the best kind we had; they were proposed a few centuries ago. But at some point, they stopped working in certain domains—say, for microscopic particles, like electrons, and protons, and so on. That's how we had to conjecture other laws, like quantum theory, for example. And now even quantum theory could turn out to be wrong! That's why physics is charming. It's fluid, never settled. It's making progress, hitting roadblocks. To make progress in physics, you need as much imagination and perceptiveness as you need to write a good story or a profound poem or to come up with a strong philosophical argument."

Rita was astonished. "I wish they had told me this from day one of physics in school. I would have taken it all more seriously."

"Yeah, sometimes teachers forget to tell you what the best bits are; sometimes they focus on the wrong things. Some teachers are all absorbed by box-ticking: do this, do that; make sure the syllabus is covered; follow the government's directives; blah, blah, blah. So they forget to tell you that there's glittering stuff and where to find it and how to see it. But the glittering stuff is there, somewhere; you just have to work a little harder to find it. I think they should allow you to be more playful with ideas and see that physics is as creative as literature or philosophy or art. It's just cast in a different language, and it focuses on things that are more removed from us, but even more fundamental."

Rita and her father paused the conversation for a while, enjoying the late-spring sunshine and the quiet all around them. As Rita was going back into the house, she paused and turned back: "One more thing, Dad. Did you say that even the current laws of physics will at some point change?"

"Yes! Yes. I wouldn't be surprised if soon there is an even stronger, more radical shift in the laws of physics than those that happened before. It's coming, I guess. We have been stuck with the same laws for a long time now, more than a century. We just have to be more imaginative, bolder, in the way we describe things."

Rita returned to her desk. Suddenly physics looked deeper. It was a tool to make sense of the universe around her. As she settled back down to her workbook, part of her mind was still enchanted by the idea of the next revolution in physics, and by the daring thought that, perhaps, even she might attempt to make it happen.

2.

Beyond Laws of Motion?

Where I explain the logic of the **traditional conception of physics**, which uses exclusively explanations by dynamical laws and supplementary conditions; why it cannot capture counterfactuals such as information, work and heat, or knowledge; and why physics needs to resort to a radically different approach, based on **counterfactuals** (statements about what is possible or impossible), to incorporate those entities in an exact and fundamental way.

There was a time, not very long ago, when our planet seemed far more mysterious than now. A vast number of things about its structure were unknown; entire continents—the Americas and Australia—were still hidden from the gaze of the rest of the world. It was the era of the great explorations by sea, when navigators set out on the oceans, attempting journeys that were more dangerous than a trip to space would be today, to find new commercial routes and knowledge and conquests. As they were trying to accomplish mundane tasks related to making their living and to expanding the trading domains of their respective countries, those explorers also made huge leaps of creativity to keep themselves safe in the watery world. They developed sophisticated knowledge about the winds and the sea currents; they constructed instruments to find the right direction during the day and at night, with the help of the stars. As they headed towards that subtle dividing line between the sky and the sea that seems like an invitation to a journey to infinity, they were hoping, ultimately, to see a less perfect profile appearing in the distance, signifying landfall after months of utter isolation.

While awaiting that sight, they had numerous fears as well: fear of storms; fear of mighty, unknown creatures; fear of sickness on board; fear of being becalmed; fear of that wavy immensity opening up all around them; fear of uncertainty. But they also had their charts, their instruments, the technology used to build their vessels. To bring in a word introduced in chapter 1, they had knowledge—information that is capable of self-perpetuation.

Knowledge allowed them to make predictions: about favourable winds and currents; about where they might encounter rocks or dangerous shallows; about how long their journey would last. Such predictions tamed some of their doubts and fears, and eased them through perils and uncertainties.

As for those early explorers, predictions are still the most sought-after output of science, and of physics in particular. They will be one of the focuses of this chapter. I shall explain the logic of the traditional way of making predictions in physics, show its limitations, and suggest how counterfactuals can remedy some of those limitations.

A prediction (in physics as well as in other fields) is a conjecture about some piece of information that is not known prior to the prediction. Like any conjecture, a prediction could be false—as one might discover by checking whether the prediction is, or is not, met in reality. False does not imply useless. An example of a false, but far from useless, prediction in maritime history is that made by Christopher Columbus in the fifteenth century. He predicted that by travelling westwards from the coasts of Europe one could reach the East—the 'Indies'. His specific prediction was, as we know, erroneous—or, to be precise, incomplete. He had not guessed that another continent was in the way. In fact, Columbus's ocean exploration is how Renaissance Europe discovered the Americas. Still, his prediction was powerful, useful, and contained some truth: had he been able to continue travelling westwards, beyond America, or thousands more miles south round Cape Horn, he would have reached India.

A notorious case of a useless prediction appears in the legend of the Cumean Sybil—the priestess who resided in the Apollonian oracle at Cumae, an ancient Greek colony where Naples is today. The story goes that a pilgrim came to her asking for a prediction about whether he would return safely from an imminent war. This was the Sybil's reply:

Beyond Laws of Motion?

Ibis redibis non morieris in bello.

The cryptic sentence contains a prediction, which is what the pilgrim was hoping for. But, unfortunately for the pilgrim, it is hopelessly vague. According to where one pauses, i.e., after '*redibis*' or after '*non*', that statement can have two completely different meanings. One is: "You will go, you will not come back, you will die at war". The other is: "You will go, you will come back, you will not die at war". Apparently, the statement was uttered only once and with a flat tone of voice—so it was impossible to tell which one of the two meanings it had.

What is the difference between Columbus's predictions and the Sybil's? The former is powerful and worthy, even if false; the latter useless. But why, exactly? The answer shall not be found by examining the content of the predictions themselves. We have to go deeper: the difference lies in what the predictions rely on. It lies in the underlying explanations. The prophecy of the Cumean Sybil did not rely on or offer an explanation of why the pilgrim would, or would not, come back from war. Without any further explanation, it is impossible to tell which one of the two opposite meanings that statement has. Columbus's prediction, instead, relied on a good explanation: that the Earth is round.

The quality of a prediction depends, ultimately, on the underlying explanation. This point is so important that we need to spend a little time reflecting on it. It is just like what happens on a long hike: when you reach a spot with a wonderful view, it is good to pause and take a little rest while contemplating the beauty of the landscape from that particular place. Our gaze now moves far away from the gloomy land of oracles and comes to rest on a boundless, shimmering prairie—a field where the connection between good explanations and powerful predictions is clear and immediate. It is the field of physics.

Predictions in physics are powerful. They supersede religious and mythological predictions and also those made by rules of thumb—rules such as "In order to grow good carrots in your garden, you need to sow carrot seeds in February". Often, laws of physics are so general that they make claims about the universe as a whole.

Take the case of Newton's laws: their explanations and the resulting predictions were intended to apply, in principle, to any system in the universe. For example, they predicted the existence of the planet Neptune, which had until then escaped all direct observations made by astronomers. Neptune's story is significant: it is a splendid illustration of how physical theories allow one to make headway into the realm of things that are yet to be known. Until 1846, the planet Neptune belonged to the vast realm of unknown things. Then, some small deviations were noticed from the orbit that Newton's laws predicted for the planet Uranus. This prompted the astronomers to conjecture that a nearby planet was possibly disturbing Uranus's motion, interacting with it via Newton's law of gravitation. Thereby, via that law, they predicted where exactly the planet, if it existed, should be observed; and when they pointed the telescope in that direction, they saw it. This story is a magisterial example of an accurate, powerful, and far-reaching prediction; most predictions in physics are of this kind. The possibility of this degree of accuracy in predictions is striking because it is hard to come by, even in many other fields of human endeavour. Just think, for example, of how difficult it is to make such predictions in medicine, politics, or in financial markets. Yet in physics that is possible.

In physics, and in science in general, both explanations and predictions must satisfy strict criteria. In particular, explanations must generate predictions that are testable. Perhaps you have noticed that being testable is itself a counterfactual property—pertaining to what can be done with the prediction. There are indeed counterfactuals at the heart of the most fundamental scientific theories and of the process of scientific discovery. Specifically, 'testable' means that it must be possible to set up an experiment to disprove the prediction, if it is false—i.e., if it does not match what is actually observed in reality. Take, for instance, a round glass marble; leave it free to move on a slide. Most of you will predict that the marble will roll down the slide, with increasing speed. This prediction is testable: if someone else said that the marble would go, say, upwards; or that it would stay where it is; or that it would start bouncing up and down, you would be able to set up an experiment to witness directly that the marble indeed rolls down. In short, you would have falsified all the other predictions by testing them against your own.

There are, of course, plenty of untestable predictions. I shall consider a hypothetical cosmology where (let me say, for the sake of amusing the reader) an immovable dog supports the universe, in perpetual equilibrium, on his head. (Incidentally, there are plenty of variants for this well-known example: you could substitute any other entity for 'dog': turtle, horse, pangolin, hippo, or whatever else takes your fancy.) The main prediction from this cosmology is that the dog is eternal and immutable; in particular, it will never turn into another animal. This prediction is clearly not testable. The dog is not accessible to us because it resides outside our universe. So, how can one even check whether it is a dog or something else? Incidentally, the dog explanation on which the prediction relies has itself several problems. For example, why should a dog rather than a turtle be supporting the universe? In the myth, that is not explained; the choice of the universe-supporting animal is arbitrary. In addition, it remains to be explained what the eternal dog rests on, and where the dog comes from in the first place. This problem is common to most religious accounts of reality. Such accounts give no explanation about why divine entities, such as our dog, exist in the first place, or about how these entities came into existence.

Why is testability of predictions so central to the progress of physics and science in general? The reason is that it provides a particularly efficient way to find mistakes in the explanations and correct them. I want to open a digression to illustrate how predictions, explanations, and testing are all intertwined within the method that allows science to make tentative progress. To this end, let me stir the cloudy water in the pond of history and bring up the spirit of the thinker who pioneered the scientific method as we know it: Galileo Galilei. Galileo's experiments to test his predictions are striking for their beauty and simplicity. He intended to test his theory's predictions about the motion of systems against those that Aristotle had proposed in antiquity, and which had been considered the authority ever since. Galileo's predictions were about the motion of a hard, smooth bronze sphere left free to roll inside an inclined smooth groove, without friction. In particular, he predicted that spheres of different sizes (masses) would undergo the same motion down the groove: the same speed, in particular. This prediction was in sharp conflict with Aristotle's theory— which predicted that spheres of different masses would roll down with different speeds. On the face of it, Aristotle's idea seems intuitively true—which makes Galileo's prediction all the more interesting.

Why did Galileo make that prediction despite its going against common sense? The answer lies, once more, in the underlying explanation.

This explanation is based on a thought experiment that Galileo invented. A thought experiment is an elegant intellectual device. It does not have to take place in reality. It just takes place in the mind, enabling us to reach some conclusion. The thought experiment in question allowed Galileo to predict that spheres of different masses should roll down the groove undergoing the same motion, with the same speed. However, the experiment itself applies to a simpler case: the motion of systems dropped from a high place and left free to fall. Aristotle's theory, as I said, was that the speed of the falling system would be greater the larger its mass. For example, Aristotle would say that a smaller bronze sphere would go down slower than a larger one when dropped from a height. To disprove this theory, Galileo first assumed Aristotle's idea was true; then, he showed that by following the logical implications of that idea, one reaches a conclusion that is in contradiction with that very assumption. As a general principle, whenever one supposes that something is true and then reaches a contradiction, one can conclude that the assumption is false. To reach the contradiction, Galileo reasoned like this: If one joins a smaller sphere to a larger sphere via a string, and then drops them both from a height, what happens, according to Aristotle, is that the smaller sphere has a smaller velocity. In this thought experiment, the smaller sphere should lag behind. If the two spheres can fall down for a long enough stretch, the smaller sphere would slow down the larger sphere, by pulling on the string. So, the combined system made of two spheres would go down at a speed that is slower than that of the larger sphere by itself. But here is the glitch: this contradicts Aristotle's idea that systems with larger masses should have larger velocities! After all, the system made of the two connected spheres has a larger mass than the larger sphere by itself. If Aristotle's idea were true, it should be faster, not slower, as we concluded! Therefore, by this thought experiment, Galileo was led to conclude that Aristotle's idea was false, and that spheres of different masses should undergo the same motion when falling freely. He then conjectured, with another leap of creativity, that they should display the same behaviour when sliding down the groove.

Galileo went on to test the latter prediction against Aristotle's ideas by observation (here is where testability of the prediction is important). He first observed what happens to the sphere when the groove is not inclined. He noticed that when the sphere is given an initial gentle push and then it is left free to move along it, it undergoes a motion that he called, in Italian, *equabile*—which means 'uniform'. In his words, that means that the sphere traverses "equal space intervals in equal time intervals". In modern physics, we would say that its velocity is constant. Galileo then observed that when the groove is inclined, the sphere undergoes a different motion. In equal time intervals, it moves by increasing space intervals; its motion is accelerated. Then, Galileo tried the same experiment with different masses and observed that the behaviour was the same for them all. All the masses were undergoing the same motion, with the same speed and acceleration, just like he had predicted. So Galileo's experiments refuted Aristotle's theory, which, as a consequence, was forever dismissed. This example illustrates how the testability of predictions in physics, and in science in general, is central for the possibility of error correction, and therefore for the progress of science as a whole. Explanations whose predictions are found wrong in an experiment automatically become problematic, and they are usually dismissed in favour of alternative explanations.

As mentioned in the first chapter, Galileo's and Newton's explanations, and the resulting predictions, have *an important trait in common*. Their approach to explaining physical reality is centered on what is usually called a *law of motion*. A law of motion, or a dynamical law, is a description of where a system (such as a sphere or a planet) goes, given that its motion starts at a certain point in space and time. Think of a sequence of snapshots, each of which represents the 'state' of that system at different times. The law of motion provides the rule for how the snapshots are ordered: in particular, there will be an initial and final snapshot, representing the starting and ending states of the motion—which in physics jargon are called 'initial conditions' and 'final conditions'. For instance, in the case of a ball fired by a cannon, the initial snapshot contains the ball sitting inside the cannon, about to be fired; the final snapshot represents it when it lands on the ground. Typically, any snapshot along the sequence is explained in terms of its predecessor—ultimately, in terms of the initial snapshot. Now, all sequences of snapshots described by known laws of motion have a particular property: each snapshot has only one predecessor and one successor in the sequence. This property is something physicists call 'reversibility' of dynamical laws: once you have gone all the way down the sequence of snapshots, you can go back without any uncertainty, because each snapshot has only one predecessor. Unlike in a garden or a labyrinth with forking paths, therefore, no bifurcations occur along the line: there is no ambiguity in how to go back or forth.

The explanation by laws of motion is the most traditional in physics. It was first introduced by Galileo; then it became established with Newton's laws; today, the two most fundamental physical theories, general relativity and quantum theory, are expressed via laws of motion, too. And so are all other theories that physicists generally consider fundamental, like those governing electromagnetic fields and elementary particles.

The long-term success of the approach by laws of motion is remarkable; its predictions are extremely powerful. Suppose, for instance, you were a general about to attack a city protected by robust walls, which you want to batter down; Newton's laws tell you exactly how to tilt the cannon in order to maximise the impact of its projectiles, by predicting their motion in every detail. For example, they tell you that there are only two possible paths available to the same point of impact for a cannonball with any particular initial speed. In both cases, the ball describes a parabola in the air, but with different maximum heights, depending on the initial condition—the angle at which the cannon is tilted initially. When the cannon is tilted at a high angle, the ball flies high and falls down beyond the walls; if the cannon is tilted at a lower angle, the ball flies lower, and it can, if the angle is right, strike the protective wall. In both cases, the description of what is going on is encapsulated in the sequence of places the ball traverses as time goes by: this set of places is the ball's trajectory—its path, which is dictated by the laws of motion, Newton's law in this case. In this approach, the explanation for why the ball hits the target at the end is given in terms of the places the ball goes through—ultimately, as I said, in terms of its initial position and velocity—the 'initial conditions' of the system's motion.

Since the dynamical-law approach is so powerful, it is natural to wonder whether it could be extended to explain everything that happens in our universe, including the whole universe itself. In other words, would a physical theory of the initial conditions of the universe and of its laws of motion provide a satisfactory explanation for everything in it? The answer, as you are about to discover, is no. I shall point out that explanations in the form of laws of motion and initial conditions are excellent for a special purpose—i.e., to make predictions about what happens on a subpart of the universe, like cannon (or tennis) balls, marbles, and planets. But they cannot explain everything in physical reality: in fact, when regarded as an explanation of everything, they have serious problems. Problems are fruitful things in physics, as they are in life. They hold the promise of improvement when they are addressed properly. These problems are the very reason why we have to venture on our journey in the land of counterfactuals.

As I said, the dynamical-law type of explanation looks like a sequence of snapshots. It has an initial and a final snapshot, and there are all the snapshots in between, whose order is set by the laws of motion. The explanation for something happening on an intermediate snapshot—for example, the cannonball suspended in the air at the highest point of its trajectory—is in terms of what happens in the snapshots before and after that particular snapshot.

Now, if the initial snapshot of the sequence reminds you of the dog in the cosmology I mentioned earlier in this chapter, you are quite right. Why should one have a particular initial snapshot and not a different one? Surely, there must be an additional explanation for that. But that explanation cannot be, itself, in the form of initial conditions and laws of motion—it cannot be given in terms of another sequence of snapshots. Otherwise, that explanation would just look like adding another sequence of snapshots to the existing sequence, placing it at the start of the latter. But that new sequence in turn would require an explanation for its own initial snapshot, and so on. In the dog-based explanation, this would correspond to supposing 'dogs all the way down', to explain the first dog.

The approach by initial conditions and laws of motion, taken in its strict form, is not a self-contained explanation for the universe. Adding sequences to the first sequence, or dogs to the first dog, does not help to address the problem. This problem is what philosophers call an 'infinite regress'.* It is exactly the same problem that religious explanations for the origin of the universe run into. How does one explain the existence of God in the first place,

---

*An infinite regress occurs when one explains something (the initial condition, or the dog) in terms of the same thing (another initial condition, or another dog), over and over again.

without assuming another god who created him? In religion, that question is left unanswered. In physics, instead, one must explain why one chooses certain particular initial conditions. The important point that we have just established is that one has to do so via some different kind of explanation. The explanation via initial conditions and laws of motion cannot be the whole story.

The issue of initial conditions is a serious open problem in physics—which has long remained unsolved. There are currently some viable proposals, which constitute the branch of physics called 'cosmology'. These theories, incidentally, are not even remotely comparable in accuracy and sophistication with other existing successful theories, such as general relativity and quantum theory. They also suffer from the impossibility of testing some of their predictions. The reason is that they are all designed to agree that the universe should look exactly as we see it now—and therefore they are all confirmed by what we see now—but it is impossible to discriminate between them by considering their predictions for how the universe should have looked at its origin because it is impossible to set up tests *then*. This does not, of course, mean that there could not be any solution to the problem of initial conditions; but currently we do not have a satisfactory one. We must therefore think of alternative ways of looking at the problem. The Science of Can and Can't provides a way because, unlike the traditional conception of physics, it does not rely on initial conditions or laws of motion as its fundamental, primitive elements.

When regarded as an explanation for the whole universe, dynamical laws are not self-contained in another, important sense. Imagine, for example, a collection of pictures of the sphere rolling down Galileo's groove, taken in rapid succession—say, one every second—to cover the whole motion. As we've seen, what a dynamical law does is put them in a particular order. For instance, supposing that the pictures were scattered in front of you, you could use the dynamical law to line them up in a row, one after the other, according to its prescription. So, you would write on the pictures 1, 2, 3, . . . , according to what the law tells you, meaning: at time 1, the ball is on the top of the slide; at time 2, it starts sliding down; and so on, until it reaches the end, at some other time, N. So, to describe an ordered sequence of snapshots, one has to refer to an external sequence—a sequence of times, say, whose elements are already ordered by labels 1, 2, 3, and so on.

We have found, again, an example of infinite regress: the same problem about ordering the scattered snapshots reappears for the sequence of N times we used to order the scattered snapshots, and so on. In general, a dynamical law must refer to some external entity—time—which is used to order all the events happening during the motion so that they do not happen all at once. Yet the existence of time is taken as axiomatic and never properly explained in terms of anything else. In addition, recall Galileo's experiment: in order to describe the motion of the sphere, he had to time it with a clock. But in the case of the universe, this constitutes a problem: What is the clock to time its evolution? The universe contains everything, by definition: there cannot be anything external to it, let alone a clock. These are two faces of the "problem of time", which affects all dynamical laws when regarded as ultimate explanations. Incidentally, this problem also affects laws as formulated in general relativity, where instead of a single external label (time), you have the set of labels specifying a point in spacetime: the same problem presents itself as far as spacetime itself, which is left unexplained. Here I do not wish to expand on the solutions to this problem. My point is just this: whatever the solution of this problem may be, it cannot be given in terms of initial conditions and dynamical laws; otherwise, it falls into infinite regress. It must be given in terms of some other kind of explanation. Some proposed explanations already exist. If you are interested in reading about them, beautiful accounts are in Julian Barbour's magisterial treatise *The End of Time* and in Michael Lockwood's intriguing book *The Labyrinth of Time*.

Surprisingly, explanation by dynamical laws also implies a large degree of arbitrariness. To see how, recall that the explanation by dynamical laws is like an ordered sequence of snapshots, where any snapshot is connected to the previous and the next one via the laws of motion; and that it is customary to assume that any intermediate snapshot is explained in terms of the initial one—the initial conditions. But why should one explain a particular snapshot in the sequence in terms of initial conditions? After all, the laws are reversible: one may well explain everything occurring in the sequence in terms of the final snapshot, going backwards from the end of the sequence, or going both forwards and backwards from any intermediate snapshot. Clearly, it is not satisfactory to have this arbitrariness regarding which snapshot to choose in order to explain the rest. But once more, the resolution must be given in terms of some different, alternative type of explanation; otherwise one falls again into the trap of the infinite regress.

I have explained a few different ways in which the traditional conception of physics is not self-contained—the problem of initial conditions and the problem of time. Now you come to a final, deep, and fascinating problem with dynamical laws, which, as you shall discover, also hints to the solution of all the others. Dynamical laws cannot handle specific counterfactual features of systems appearing in our universe. They cannot express them fully and adequately.

First there is the kind of counterfactual that declares some transformation to be *possible*. Consider a specific transformation—addition: x and y (two numbers encoded in some numbering system) must be turned into the number $x+y$. When we try to express the fact that addition is possible in the dynamical-law approach, we encounter a number of subtle and important problems.

One way to express that addition is possible is to say that an adder is possible. An ideal adder is a machine that, when given any two numbers x and y in input, gives in output $x+y$, and—mind you—it remains unchanged in its ability to do that again, with other numbers. The ability to work in a cycle guarantees that the adder can add again, if needed. An approximate adder is included in any smartphone calculator. I say approximate because after a certain number of years, the smartphone's ability to add will wear out, and the precision of addition will deteriorate inevitably. Also, the input will be encoded in a limited number of digits, hence achieving only a limited precision.

The possibility of an adder cannot be expressed fully if one wants to explain everything in the universe using only laws of motion and a given initial condition, as in the traditional conception of physics. For a start, by fixing a specific initial condition, the universe evolves along a single, particular trajectory (set by the said initial condition). No ideal adder will ever appear on that trajectory! On that trajectory, there will only be processes implementing approximate adders, with limited precision, where only a fixed, finite set of inputs is ever added before the approximate adder wears out. Any particular instance of an approximate adder lasts only a finite time and will ever only add a given sequence of input numbers. (Otherwise, we would have a violation of the condition that the laws are no-design, as I explained in the first chapter.) That an adder is possible, or that addition is possible, means far more than that. First, it means that the adder, when given as input any two numbers, can output their sum. This fact refers to any two numbers; never mind whether they are actually given as input to it in reality once the trajectory is selected. Second, that an adder is possible means that there is no limitation to how well it can be approximated by any of the approximate adders. But this fact, once more, cannot be expressed within the traditional conception of physics—because the latter can at most say what happens to a particular instance of an approximate adder, if it occurs on the trajectory of the universe. It is going to be true on each of the possible trajectories, for each of the allowed initial conditions. So even the enumeration of all possible scenarios that would happen if a given initial condition were to be set would not express the possibility of an adder, either.

Another type of counterfactual that cannot be accommodated in the dynamical-law approach is the fact that something is impossible. Think of the principle of conservation of energy, which tells us that a perpetual motion machine is impossible. In the dynamical-law type of approach, one can only say that a perpetual motion machine does not happen: that means that no point on the trajectory of the universe contains one, given a particular initial condition. But that a perpetual motion machine is impossible does not mean it does not happen under a particular initial condition! It means it cannot be built under any of the initial conditions and any of the actual dynamical laws. This statement is much more powerful, and categorical, than any of the statements one can make about what happens on a particular trajectory.

One might try to express the fact that something is impossible by stating that under no initial conditions would that particular thing happen. This would, indeed, be a law; but that law does not completely cover the meaning of a principle saying that something (e.g., a perpetual motion machine) is impossible, because it holds only for a particular dynamical law; whereas principles about impossibility hold, as I said, for any dynamical law. They are much more far-reaching.

A final problem about the dynamical-law approach is that it seems, on the face of it, to conflict with the existence of entities that are capable of making choices—such as you and me. Every omniscient narrator knows this. The omniscient narrator is the entity that tells the story in a novel, in third person. The narrator knows all the thoughts of the characters in advance. Their ideas are set from the very beginning of the novel. Choices only look like true choices to the characters, but in fact they are predetermined and fixed by the narrator's plans. Likewise, the explanation based in laws of motion and initial conditions would seem to imply that so must be ours. Our choices, and everything else depending on them, seem already to have been set in advance; they are written in the dynamical-law explanation, and fixed by the initial condition of the universe. The dynamical laws' sequence of events fixes everything—it is given once and for all. All your ideas are laid out there: there seems to be no room for them to be unpredictable, as they should be if they were truly 'free choices'.

We have just outlined what is called a 'deterministic nightmare': the fact that there does not seem to be any room for free choice if one presupposes the existence of a fixed, predetermined story for our universe, which is the picture that the traditional conception of physics, in terms of dynamical laws and initial conditions, seems to suggest. For example, whether tomorrow you will have croissants or kippers for breakfast has been fixed at the start of the universe, in its initial conditions. The same goes for the fact that you are reading this text right now, instead of some other book (or maybe doing something else altogether, such as watching your favourite show), all determined at the beginning of the universe down to the precise words, typos included. Unpredictability of action, or free will, is therefore another counterfactual that the dynamical-law approach does not seem able to accommodate. We do not yet know how to accommodate exactly free will in physics—but that only means we have to think harder. This problem exists, but it is not insoluble. It only appears to be so if contemplated from the narrow dynamical-law type of approach.

Fortunately, the dynamical-law approach is not the only way to provide explanations and predictions. Why should all good explanations look like chronologically ordered stories which unfold from beginning to end? The fact that something has happened before something else need not be the whole explanation for how systems work in physical reality. To see more clearly how else one can provide explanations, let's imagine a chessboard and our generous storyteller (whom we encountered already in chapter 1), who will tell us a story about it.

Chess is a beautifully rich game and a great source of inspiration. I often contemplate the pieces arranged on either side of the board, waiting to be set into motion. It is the only battlefield where a bishop and a rook are part of an army.

I would like you to imagine a draw—a situation where it is impossible for either player to win. In particular, that means that the board and the pieces are in a state such that no valid move can cause a checkmate. (Checkmate is when the king is in a position where it cannot escape capture.)

There are different types of draws; but here we can choose a particular draw, where the game ends with a single king on one side and a king and a knight on the other side. Now, there are at least two different stories, or explanations, for why a particular configuration is a draw. One is a story about the initial conditions and dynamical laws of the whole universe: this story determines, given a particular initial condition of the universe, the states of the minds of the two players and says that whatever their next move is—whatever their strategies—according to the laws of motion and the initial conditions, there is no way for that configuration (one knight and one king on one side and one king on the other side) to progress to a new configuration where there is a checkmate. This explanation is the only kind of explanation for the draw that the dynamical-law approach permits. It is very expensive, as it requires computing all the interactions between the particles that ultimately constitute the brains of the players and the chessboard; and it requires us to do so for all the possible initial conditions of all the particles in the universe that are relevant for the state of the brains and of the chessboard. Even then, it misses the point of explaining the draw.

The real explanation for the draw is that, according to the rules of chess, the knight and the king can move only in certain prescribed ways; specifically, for them, only some moves are possible, and others are impossible. This explanation, invoking counterfactuals, is far more exhaustive and far more specific. In addition, it does not need to include anything about the rest of the universe and the minds of the players; in fact, it involves only the counterfactual properties of a small portion of the board and the three pieces left on it. Finally, the explanation can also be used to make predictions, because it allows one to say that the configuration of the board will not evolve into a checkmate configuration, provided that the players stick to the rules of chess. However, this explanation does not look like a story unfolding in time: it says simply that some transformations are possible, and others are not; it is not given as a sequence of moves, or more generally, it is not a dynamical law. It is a story of a different kind: it is about the impossibility of transformations on the pieces on the board. It is independent of time; and it requires contemplating not just things that happen, but also things that can and cannot happen.

Can this logic, adopting counterfactuals, be fruitful in physics, other than in my elementary example? Indeed, in physics and science in general, we already resort to modes of explanation other than dynamical laws, some of them adopting counterfactuals. What we have in place is, in fact, a hybrid approach. For example, physics resorts to principles like the conservation of energy, which, as I said, are about counterfactuals, too. These principles are not in the form of dynamical laws: they are statements that require certain things to be impossible, such as perpetual motion machines. Yet they can be as powerful as dynamical laws at generating predictions. A famous example is the prediction of the existence of the neutrino—a previously unknown elementary particle. This prediction is akin to the prediction of Neptune's existence—but this time, it was a prediction of the existence of a subatomic particle, not a planet; and the prediction was generated from a principle, not a dynamical law. The prediction was obtained by reasoning that without that particle the law of conservation of energy would be violated. It couldn't have been obtained from a dynamical law, because the laws of motion of neutrinos were not known until much later! Principles appear also in Newton's laws. Only the second law is a dynamical law—relating the force on a system to acceleration and mass; but the other two laws are not really dynamical in their existing formulation. The first law is a hybrid one. It says it is impossible for a system to change its state of motion when it is not acted upon by any force; hence, the system will continue in its given state of motion until some force intervenes. The law does refer to dynamical laws, via the concept of 'state of motion'; but it mandates (like a principle) that some transformations are impossible—

specifically, those changing the state of motion of a system without it being acted upon by a force. The third law is even closer to being a pure principle: informally, it requires that to every action there must correspond an opposite (in direction) and equal (in magnitude) reaction. If, while on a walk in the park, your dog is pulling you ahead via the leash, you are pulling the dog back with an equal and opposite pull. This fact is necessitated by the principle included in the third law, not by a specific dynamical law.

Wouldn't it be wonderful if it were possible to take inspiration from these principles, which relate to counterfactuals, and imagine an entirely different way to formulate the laws of physics, one that takes counterfactuals as primitive and the laws of motion and initial conditions as derivative? One could even conjecture that this new mode might solve the open problems in the dynamical-law approach, as well as fill in the gaps of existing theories, while still covering all their predictions.

The kind of explanation I am imagining, which is that provided by the Science of Can and Can't, is even more radical than the hybrid type of explanation which we currently use in physics. It places counterfactuals at the most fundamental level; then it explains dynamical laws and initial conditions in terms of them. These laws of motion can then be used to make testable predictions about cannonballs and electrons, much as they are now. But their underlying explanations would be in terms of principles about counterfactuals; and this could provide a solution to the infinite-regress type of problems I mentioned earlier (e.g., resorting to an infinite set of initial conditions). Just as the counterfactual properties of the chess pieces (about what moves are possible and what are impossible) can explain the draw on the chessboard, so counterfactual properties can explain why the universe is in a certain state, avoiding mention of the initial conditions altogether. Both statements (about possibility and impossibility) are equally important: you will see several examples of laws about possibility and impossibility in the chapters to come.

These may seem like bold speculations; and they are. The first time I encountered the idea of reformulating physics with counterfactuals was in a proposal by the physicist David Deutsch. At the time, I thought it was fascinating, but crazy. That was during my doctoral studies in Oxford, when (to put it as Alice in Wonderland would) I started trying to imagine "as many as six impossible things before breakfast". That idea was one of them. But within a few months, David and I were working together on a paper developing this idea and applying it to information theory; and after my doctorate, I decided to focus completely on pushing it further to try to address various unsolved problems in physics. By then, I was convinced its promise was enormous. My research work to date, with the help of a few brave students and a handful of other physicists, has concentrated on putting this approach to the test. In the following chapters, I shall explore the problems that this approach has solved so far, and its potential to solve further problems. It's time to journey deeper into the land of counterfactuals.

La Locanda della Grotta

Apart from the rowdy congregation of physicists sprawled around the long wooden table laden with glasses and bottles of beer, port, and cider, the garden of La Locanda della Grotta was almost deserted. The only other customer was a lone man sitting in a corner. He was smoking a cigar, in silence, and his figure, elegantly dressed in a dark shirt and jeans, was enveloped in aromatic swirls of smoke. He seemed lost in his thoughts, uninterested in what was going on around him.

A few tables away, the innkeeper was eagerly cleaning up in preparation for the next day. Her young son was behind the counter, pretending to help. In fact, he was mostly playing, and *also capturing snippets of the conversation. He and his mother knew enough English to roughly follow the discussion, even though the technical physics terms were beyond them. They were both curious about the weird crowd that had descended on the otherwise utterly quiet town of Cortona.*

The scientists were there for the annual conference organised by the Scuola Normale Superiore di Pisa, and apparently, this year, the topic was thermodynamics. Professor Flagg, a prominent physicist who had just delivered the main colloquium that evening, sat at the head of the table like an almighty god on his golden throne, surrounded by minor deities (his colleagues) and ordinary mortals (the students) in adoration.

"Thermodynamics is a worthless subject", Professor Flagg was pontificating. "Why are we even having a conference about it? Everything is in the dynamical laws. We don't need more— heat engines and computers and all that—they all just boil down to the dynamical laws of a bunch of particles. Yes—young lad from the back! Do you want to ask a question?" One of the local high school students, invited to the conference to encourage their particular interest in science, timidly raised his hand and asked Flagg something about entropy. "Do yourself a favour", came the scornful reply. "Go and read my book. It's a must for someone of your age. It will enlighten you." A pause; another glug of port; then Flagg turned to another student: "Next question?"

The innkeeper shot Flagg a disapproving look over her shoulder. He was insufferable. His pompous demeanour, his conde-

scending manners, his arrogance and tone of superiority were out of place. He was supposed to help young kids understand more about physics, not advertise his own status. But in the end, it was not her business. She glanced over at her son, who seemed to have lost interest in the conversation and was completely absorbed in his games. "Good", she thought. "At least he will not pick up that guy's horrible manners."

The evening went on in this way for a while: the students kept asking questions, and Flagg replied along the same repetitious lines.

At one point, a female student raised her hand. She was the only girl at the table. The innkeeper sharpened her ears.

"Do you think that principles in general, like the principles of thermodynamics, could sometimes guide our search for dynamical laws? I have read that Einstein used this principle-based approach to derive his theory of gravity. How do you reconcile this with your views?"

Flagg made a face, and then looked at her with a pitying expression. "Heavens, what a silly question! Of course Einstein concocted some kind of narrative for how he had arrived at relativity. But ultimately, what he found were laws of motion. So the principles were irrelevant to physics, just like all principles. In fact, he later admitted that his principle-based approach was highly misguided. I don't think this in any way contradicts my view. If you want to be a physicist, stick to the equations, my girl, and don't trouble yourself with crackpot philosophy."

The student sank back in her seat with a flustered expression on her face, and an awkward silence fell around the table. After all, what she had asked made sense, but no one dared to challenge Flagg. He was too almighty to be contradicted or criticised.

At that point of maximal embarrassment, a voice from the back of the garden broke the silence. "I have a question, too." The solitary man, who had been sitting and smoking in silence the whole time, ground the last smouldering wad of his cigar into the ashtray with a slow, circular gesture. Swirls of smoke rose up in the air, emanating a subtle aroma of leather and oak wood. "I would like to know what use is a physicist who cannot argue", the man continued calmly.

Silence.

"You have been rambling around, Professor Flagg, but you haven't really replied to the question you were asked."

Flagg had been taken aback, but was now regaining his pomp-ous air. "I'm afraid we have not been introduced. You are . . . ?"

"It doesn't matter who I am, really. I'm just pointing out that this young woman's question has remained unanswered." The mysterious man was sitting in a seemingly relaxed posture, but his sharp blue eyes transfixed Professor Flagg with an uncom-mon intensity.

Flagg felt an unaccustomed need to justify himself. He swal-lowed hard. "Er . . . well, I . . . I did answer the girl's question!"

"You did not", the man replied in a flat tone. "Your answer was not to the point. Let me sketch what a complete answer could

look like. A complete, honest answer might say that physics has so far operated with a mixture of principles and dynamical laws, and that Einstein, just like you or anyone who does physics, also worked with that mixed approach. A bit of principles, a bit of dynamical laws. The dynamical laws really don't tell you the whole story; one, because they may not be the ultimate dynamical laws; two, because it is impractical to apply them to large aggregates of matter; and three, because we do not even know the initial conditions. Yes, principles supplement them—and in a way that is powerful and unexpected. Einstein had understood that very well. And the young woman's question, which was far from silly, pointed that out very appropriately."

The innkeeper relished Flagg's look of dismay. It was clear by then that the man with the cigar knew a lot about physics, and he was doing a better job of answering the student's question than the illustrious professor.

"OK, thank you for sharing", Flagg finally said. There was a nervous tinge in his voice. "Are you finished now? This is a private discussion."

"No, I'm not finished. I am just getting warmed up. I'm surprised that someone who presents himself in his books as a passionate scientist and a champion of physics is not acquainted with the basic practice of proper scientific discussion. I will explain the point. You think that reductionism is the basic principle (funny, no?) of science? Fine. Defend that idea with your best arguments; and if someone asks you a question that you don't have

the answer to, don't be afraid of saying so. This is not a job interview, nor an advertisement for your books. You are discussing science here! And these young students are the scientists of tomorrow. Talking to them is one of the most important things you could be doing, aside from physics itself. Embrace them, don't dismiss them. Don't let them down. They have come here expecting answers that make sense. Maybe one day they will think of a new law of thermodynamics or of a new law of motion. Who else will?"

While Flagg and his audience were still speechless, the mysterious man stood up slowly, put some money on a nearby table to pay for his drink, and walked out. His tall and slender figure fluidly merged with the darkness outside the gate, disappearing from sight.

A little later, the innkeeper was tucking Luca into bed. "You should have gone to bed much earlier, Luca", she said.

He smiled.

"What?" she asked.

"Mamma, I really like that man."

"Who?"

"The guy who stood up for that student, earlier."

"Oh, that guy. Yeah—he gave that pompous idiot what he deserved."

"Who is he? I see him at the inn occasionally, but I've never tried to talk to him; he always looks a bit aloof."

"Oh, we do not know much about him. He lives in the hills with his wife; they have a big mansion. I think they were both successful physicists somewhere abroad. Then they left the field, suddenly. He once told me that they were both disillusioned with the whole academic enterprise, so he and his wife just left. Now they run a school for kids who want to learn physics. I do not know exactly how it works, but it sounds like a good thing."

As he settled down under the covers, Luca was still thinking about the mysterious physicist. He really liked him. "He looks like a pirate", he thought. "One of the good kind, who fight for noble and righteous causes. Maybe all good physicists are like that."

A few minutes later, the boy fell asleep with a smile on his face, having decided that he wanted to be exactly like that when he grew up.

3

Information

Where I explain how information can be completely captured within physics with two counterfactuals: the possibility of **copying** and of **flipping**; where you encounter the counterfactual property of **universality** and learn how it enables **universal computers**.

When the night falls on Sentosa—a small satellite island of Singapore—a remarkable spectacle takes place. The best location to witness it is somewhere along the bridge connecting Sentosa to Singapore. That bridge is made of smooth wooden tiles; it also has a panoramic spot where one can rest, leaning on a balustrade, while enjoying the view that spans the whole bay.

As the twilight turns into night, distinctive kinds of lights and sounds fill the air. The surroundings gradually become darker and darker, until the whole backdrop is pitch black—both the sky and the sea.

At that point, the spectacle reaches its peak. Suspended several dozen metres above the sea, green-lit cable cars run smoothly through the air, back and forth, in constant gentle motion; boats and ships move lazily across the bay, their signalling lights cutting through the darkness; the music from the bars on the shore spreads around, whispering in the warm equatorial night. Farther back, in the distance, a lighthouse flashes, on and off, on and off.

The objects populating that nocturnal landscape display extremely diverse behaviours, each explained by a different branch of physics. Cable cars, boats, and ships are powered by engines, explained by laws of thermodynamics. The music and its propagation are explained by the theory of sound. Sound is composed of waves of molecules of air, which travel to one's ears and are then converted into electrochemical signals in the brain. Molecules are in turn composed of atoms, and atoms are made of subatomic particles, such as protons, electrons, and neutrons. Light is explained by the laws of electromagnetism—Maxwell's equations. At the most fundamental level, all those phenomena are explained by quantum theory and general relativity—the two deepest explanations of physical reality we possess at present.

Despite being so different in their specific details, those systems have something in common, which is not explained by any of the existing branches of physics. The music, the boats' lights, the lighthouse—they are all signals. They are *capable of carrying information*. This property is a key trait they all share, one that (contrary to what one might think) is possessed only by a particular class of systems in our universe. (In a short while, I shall give you examples of systems that cannot carry information.)

What is the property that makes those systems, and many others like them, capable of carrying information? Answering this question involves counterfactuals; and it will keep us occupied in this chapter. It will reveal the way to express information as a fundamental entity in physics, and the fundamental physical laws that rule the physical system capable of instantiating information. This is crucially important, not only from the point of view of understanding the universe in a deeper way, but also because information and its connection with physics is at the heart of information technologies that could further revolutionise our civilisation, such as the universal quantum computer that we will discuss in chapter 4.

At first, it may seem that information does not really have anything to do with physics. In fact, in everyday language, the word 'information' is used to refer to all sorts of things. For instance, many systems contain information: books, newspapers, and magazines; emails and messages; the words we utter when speaking to friends and family; poems, songs, and ballads. The biosphere contains information, too, as I mentioned in chapter 1, encoded in DNA molecules. But even though all those systems are plainly part of physical reality, it is hard to identify information with a particular physical system. Information looks more like an abstract entity, and it is hard to pin down its connection with physics.

For a start, information does not have a specific embodiment, but it can be embodied by many diverse physical systems. Is it, then, some kind of property that systems have, such as, say, colour? That a laser light is 'green', for instance, means that the photons it emits—the quanta of energy it is composed of—have a particular frequency or energy content. Could information be something like energy or frequency? Not quite. Those are factual properties— because they are specified exclusively by the system's state at a certain time and in a certain location in space. But when it comes to information, the story is different. As I shall explain in detail, one cannot say that some system contains information just by stating a full description of its state and its factual properties— because the fact that it does has to do with certain transformations being *possible* on it. Then what do we mean when we say that a computer or a smartphone is 'carrying information'?

Rather than seeking to define information as a physical entity, or as a property of physical systems, the key is to change our focus and ask a slightly different question: What is different in the state of affairs of a physical system between when it does carry information and when it doesn't? It is, as I shall explain, a set of *counterfactual properties*. Once we pin those down, we will have established what is required of a system in order for it to contain information, and the connection between information and physics, without ever actually having to define information directly.

Let's use a thought experiment to identify the counterfactual properties that a system must have to carry information. First, we take a system that *can* carry information—such as a lamp that can be used to signal—then we gradually subtract its key properties until it stops being capable of doing so. At that point, we will know that the properties we removed in our thought experiment are necessary for carrying information.

Suppose that you were standing on the Sentosa bridge's viewing point, at night, and you had the task to communicate with an approaching boat using a lamp. The lamp is green, say, and it can be switched ON or OFF; and the code of communication is that if the lamp is ON, then the boat can proceed; if it's OFF, then the boat should stop.

Now, imagine the colour of the lamp changed. Clearly, that would not modify its ability to convey the signal. Nor would changing its shape, or similar other properties of its state. But suppose now you modified some of its functionalities. For instance, suppose that once you switched the lamp ON, it could no longer be switched OFF. Would this lamp work as a signal? Clearly no. Given that it *cannot be in any other state*, it is useless for signalling one of two alternatives: it has only one state available.

Now imagine you wrapped the lamp in a completely opaque covering, which does not allow the light to come through and to be seen at a distance. With this modification, the lamp would not be able to signal either, because it *could not be seen* from the approaching boat.

This example offers a lesson you can then generalise: the fact the light, when it is ON, carries information is due to the fact that *it could be set to a different value*, OFF, and that the difference between ON and OFF *can be perceived* by the approaching boat. Both of these properties of the lamp are counterfactuals. Abstracting from the example, you can assume or perceive a general, fundamental regularity in nature: any system containing information must have these two properties.

Property One is that *it can be set to one of at least two states.* For example, if it has two possible states—let's call them 0 and 1, generalising ON and OFF—these two states can be changed from one into another, like this:

$$1 \rightarrow 0$$
$$0 \rightarrow 1$$

This notation specifies the following transformation (or task): *If* given 1, turn it into 0. *If* given 0, turn it into 1. A machine can perform this task if it can indeed obey *both* these requests. I shall call this transformation a 'flip' (a special case of a permutation), but if you are familiar with computer science, you will know that in the jargon of that field it is called a 'NOT operation'. The name could not be more appropriate: the operation describes exactly the behaviour of someone with a contrarian personality. If you say yes, they will always flip your statement to its negation and say no; and vice versa. Likewise, this operation flips the state of a system to 0 if it is in 1, and to 1 if it is 0. We've already seen the flip operation appear in several places in that view from the Sentosa bridge. It is in the lighthouse, whose lamp flips in its ON-OFF-ON-OFF pattern; it is in the signalling lights from the boats, which also operate like switches. It is realised, to a high degree of accuracy, in any computer, when a transistor switches on and off; it is even realised, to a lower degree of accuracy, in our brain, when a neuron fires and then becomes quiet again. And, as we have seen from the lamp example, it is the counterfactual property that is necessary to send the most elementary signal—a binary one.

Property Two, required for some system to contain information, is that its states (e.g., the ON and OFF states of the lamp on the bridge) *can be received and distinguished* in some other location (e.g., by the boat's communication system). This property is trickier to express; still, it can be elegantly and fully captured by counterfactuals. It is the possibility of performing a *copy-like operation*. Remember that I already mentioned replication in chapter 1, which is a special case of copying. To see what copying is, we can dramatise the communication between the bridge and the boat by adding further layers of communication. Imagine that it is a foggy night, and that even the strongest lamp can be seen at no more than 500 metres from the bridge; but the requirement is that it can be seen by boats that are 1 kilometre from the bridge, along a particular straight path joining the bridge and the entrance of the harbour. One way to deal with this problem is for another boat to anchor at about 500 metres from the bridge, along that path. If that boat has another lamp, which can in turn be seen from other boats approaching, it can signal to them, by setting its lamp to ON or OFF, in coordination with the lamp on the bridge (just like old-fashioned telegraph or beacon signalling). The communication is successful if, whenever the ON state appears on the bridge, the boat sets its lamp to ON, too; and if, whenever the state of the lamp is OFF on the bridge, the lamp on the boat is also set to OFF. This process amounts to *copying* faithfully the state of the lamp on the bridge onto the lamp on the boat: the state of the lamp on the bridge is perfectly reproduced by the lamp on the boat.

In general, a copy-like operation is a transformation that transfers faithfully the value originally held by some system (the lamp on the bridge) onto another system (the lamp on the boat) while preserving the former's original value, too. It can be expressed like this:

$$10 \rightarrow 11$$
$$00 \rightarrow 00$$

I am using the digit in the first position on the left to specify the state (1 or 0) of the system holding information to be transmitted (e.g., the lamp on the bridge) and the digit occupying the second position on the left to specify the state '0' of the system that has to receive the information (the lamp on the boat). The overall expression with arrows specifies a transformation: IF given 10, produce 11 in output; IF given 00, leave it alone. That transformation is a 'copy' because, after performing it, the second system (whose state is represented by the digit on the right) contains the value (0 or 1) that the first system held before performing the copy, while the first system is unchanged.

In our Sentosa landscape, the lighthouse is capable of undergoing copy-like operations because its signal can be copied into the lamp sitting on a boat and then transmitted to another boat, and so on. The copy operation is also central in other systems that can contain information: a DNA molecule, as mentioned in chapter 1, is copiable in exactly this sense: the genes can be faithfully replicated into a new DNA molecule when replication occurs. The printing of newspapers from the pattern of movable type produced in the press office is a copy operation; the phenomenon by which the news gets into our brains is yet another, from the page of the newspaper to our memory. Similarly, all the signals appearing in the panorama from the Sentosa bridge have the property that they can be copied into our brains when we contemplate that view. My recollection and description of the view is a sort of copy of it (perhaps with a number of small ornaments I added to amuse you, as all writers do; and a number of small memory imperfections). That recollection is possible because the landscape contains information itself—it contains signals, which can flip (as I said for the lamp) and that can be copied in my brain and later retrieved. The copy operation is also essential for the inner workings of a computer: whenever the output of a computation is ready, it must be possible to copy it onto some other medium—for example, another portion of memory—where it can be processed further.

We have reached an important conclusion. A physical system is capable of carrying information if it has these two counterfactual properties:

1. It *can be set* to any of at least two states. (The flip operation is possible, under the laws of physics.)

2. Each of those states *can be copied*. (The copy operation is possible, under the laws of physics.)

So here is the reason why 'information' is a physical property: whether or not some system carries information depends on whether the laws of physics allow for these two transformations on that system. If they don't, then the system cannot carry information. In a universe where no system had both properties, information would not exist. So whether or not information is permitted depends on whether the laws of physics permit certain kinds of counterfactuals. But it is not a property like having a certain colour or mass—factual properties of a system. It is a counterfactual property, because whether a system contains information or not depends on whether those two transformations *can be* realised on it. Through counterfactuals, you have arrived at the elusive connection between *information* and *physics*!

Systems with those two properties are 'information media': all information media, despite their differences, have in common the fact that those two transformations are possible on them. All the systems I mentioned in the Sentosa example are information media: they have those two counterfactual properties. The simplest information medium—the fundamental unit of information—is a *bit*: it is an information medium with two possible states, 0 and 1. Its *capacity* is that it can signal at most *two* different messages. You can think of countless ways in which our universe can embody a bit: the lamp of our previous example, which can be ON or OFF; an arrow that can point up or down; a coin resting on a table, which can show heads or tails; your answer to a yes-or-no question, and so on. Thinking in terms of information allows one to forget about all the differences in the physical details of those systems, and consider them all as the same thing: a bit. The same holds for information media with higher capacity (those that can hold more than two messages)—they, too, can be thought of as made of bits.

But not every system is an information medium. A good example is a memory in a computer that is full but cannot be erased: it is possible to read information *out*, but not write new information *in* (because no more space is available, and reset is not possible). It was an information medium once, but no longer. You could also have a case where information can be copied *in*, but not *out*. Have you ever tried to write something on the foam on top of a cappuccino or a beer? At first it looks possible. But the letters rapidly fade away, to the point that they can no longer be read. Neither of these two types of systems would be capable of carrying information, because they do not have enough counterfactual properties. They are not information media.

One of the most striking properties of information media is that in that regard they are all interchangeable, because information can be *copied* from one to the other *irrespective of their physical details*. I shall call this property—the possibility to copy information from any information medium to any other—*interoperability*. For example, the information in a bit can be copied into any other bit irrespective of what physical system it is—a transistor, an arrow, a coin, or a switch. The music that has been recorded on old vinyl discs can be converted and copied into digitally encoded music on a flash memory; the sound produced by our voice can be turned into words stored in the transistors that compose the memory of our smartphone, via voice recording; the thoughts in my head can now be faithfully copied on this page; they will then be copied into your brain, and then, possibly, copied further into other brains or your notebook, if you decide to write them down. All these information media are interchangeable, or interoperable, and information can travel among any of them without restriction.

Interoperability is due to the fact that all information media have in common properties—the two counterfactuals I mentioned above—that transcend most of their specific details (i.e., whether they are photons, transistors, the spins of an electron, neurons, or switches in a lamp). In all these cases, when interested in the information-processing abilities of these systems, we can abstract away their irrelevant details and simply talk about them as information media, considering their 'information-carrying attributes' only (e.g., up/down for an arrow, on/off for a lamp, and so on).

Now, armed with these counterfactuals, you can understand why information looks like an abstraction and yet is grounded in (counterfactual) physical properties. When talking of a bit, we do not need to mention what physical system embodies it. What matters is that a bit is an information medium—entirely defined by its counterfactual properties, which hold irrespective of its physical details. What is the connection with physics, then? The key is that which physical systems are information media and which are not is established precisely by the physical laws that rule our universe. And the interoperability of information media is a counterfactual property of physical systems: it is a property of the physical world, just like the colour of the summer sky, the shape of rainbows, or the attractive interaction that holds between charges of opposite signs.

Time and again, you will see that seemingly innocuous and simple properties (such as interoperability) have far-reaching consequences. In this chapter I want to explore one of these far-reaching consequences of interoperability: the possibility of copying information from any physical support to any other is necessary for *computers* to exist (and all the related information technologies).

Because computers rely on the counterfactual properties of information media in a fascinating and subtle way, I want to open a digression about computers. In certain circles, computers have an unfairly poor reputation. In traditional physics, for instance, they are regarded as 'emergent', 'macroscopic' systems, thus not at all fundamental or worthy of attention, unlike, say, elementary particles. In the general culture, they are regarded as having interest only within a niche made up of 'geeks' (like the character Sheldon in *The Big Bang Theory*); at best, they are perceived as useful but dull machines that are only slightly more interesting to us than a dishwasher or a vacuum cleaner. In fact, we would not be considering them at all were it not for the fact that they happen to be useful for typing up essays, playing games, writing emails, shopping, dating, and socialising.

All these views are inadequate because they disregard the fundamental aspect of a computer: the intimate connection between computers and the laws of physics, which goes via counterfactuals. Let me start with the link between computers and physics. Computers are embodied in physical supports—they are made of information media (typically, billions of switches or transistors). Therefore, they are ruled by the laws of physics. In particular, which *computations* a computer can or cannot perform depends on what the laws of physics permit. This connection between computation and physics was not fully understood until the 1980s, with some of the pioneers of quantum computers. It was hinted at by imaginative thinkers, such as Rolf Landauer, Paul Benioff, and Richard Feynman; but it was fully expressed for the first time by David Deutsch and further developed by the masterful computer scientist Charles Bennett.

A simple example of a computation is the addition of two numbers (which you encountered in chapter 2): its inputs are the numbers x and y (e.g., 5 and 10), and the output is the number $x+y$

(e.g., 15). That a computer is *capable* of performing a computation such as addition means that every time it is given the right input (the two numbers x and y), it is supposed to provide the desired output (the number $x+y$); and it can do that over and over again.

The set of all computations a computer is capable of performing is its *repertoire*. So, for example, a calculator is a computer that has addition, multiplication, subtraction, and division in its repertoire.

What decides the repertoire of a computer? The physical laws that rule its components. Under given laws of physics, for each computation that is physically possible, at least one kind of computer is capable of performing it. By 'computer' here, I'm not necessarily referring to something as sophisticated as your personal computer. I mean a special-purpose computer, which has only a few computations in its repertoire—for example, the adder mentioned above, which can, as I said, output the number $x+y$, given two numbers, x and y, in input; or a multiplier that, when given x and y in input, provides $x \times y$ in output.

How do we get from a computer with one repertoire to one that has a larger repertoire? For instance, suppose you have an adder and a separate multiplier. How do you get to a computer that can perform both addition and multiplication? Under the laws of physics of our universe, thanks to the interoperability property, there is a straightforward way. All you need is a third computer that, when given in input the numbers x and y and the command 'Add', sends them to the adder; whereas when it receives x and y

and the command 'Multiply', sends those two inputs to the multiplier. Together, these three entities form a more general computer that can perform *both* addition and multiplication.

Proceeding in this fashion, nothing stops you from imagining a computer that has *all* the physically possible computations in its repertoire. It is a *universal computer*: it can be programmed to perform any calculation that is physically allowed by certain physical laws. It so happens that the laws of physics of our universe do not forbid a universal computer. Computers such as our laptops and personal computers are universal in this sense.

Another fundamental trait of computers in our universe is that all the computations in their repertoires can be realised by combining a smaller number of basic computations—which work like letters of an alphabet to compose words. This, too, is a peculiar feature that holds in our universe, but need not hold in general.

For example, 3 is a number, 4 is a number, and if we juxtapose 3 and 4, we find another number, 34. Any number can be represented, in the decimal basis, by juxtaposition of the numbers from 0 up to 9. Likewise, elementary computations can be composed with one another to realise all the computations permitted by the laws of physics. For instance, suppose you perform the *flip* twice on a bit: you see that if the bit is initially 0, it is flipped to 1; and by applying the flip a second time, you obtain 0 again! Likewise, if the bit is initially 1, after two flips it gets back to the state 1. So applying the *flip* twice to the same system corresponds to performing a different operation—in this case, doing nothing, or leaving the bit alone. A set of computations that, composed with one another, permit one to recover the whole set of possible computations in the repertoire of the universal computer is called a 'universal set'. When there is a universal set, any computation is reducible to a sequence of elementary computations selected from the universal set. These elementary computations are, in this respect, a bit like LEGO bricks: anything that is allowed in a LEGO world (from cars, to villas, to pirate ships) can be decomposed into elementary LEGO bricks of a few elementary different kinds, whose basic composition rules are fixed. Likewise, when there is a universal set, any physically allowed computation can be decomposed into a set of elementary computations from the universal set, sometimes referred to as 'gates', which can be composed according to fixed laws. When the laws of physics say that a universal set of computations is *possible*, we say that they display 'universality'. Universality is a counterfactual property (about what is possible), and it has sweeping consequences: it is universality that permits the existence of a universal computer, like the ones we use nowadays. This property was first grasped in the Victorian era. At that time, the inventor Charles Babbage proposed a scheme to build what he called the Analytical Engine. This would have been, if realised, the first programmable computer—the ancestor of our modern ones, only far larger and made of brass mechanical cogs and wheels. Ada Lovelace, Babbage's collaborator and a brilliant mathematician, understood the universality of this machine, conjecturing in her notes that the Analytical Engine could be used to produce all sorts of information-theoretic outputs, not just to compute functions. She even speculated that it could be used to produce sophisticated music. Unfortunately, Babbage's idea was not realised in practice for a lack of funding; and the property of universality was not studied until much later. It was Alan Turing, with his computing machine, who finally formalised the idea of universality in the 1940s. This concept was then sharpened and connected to physics by David Deutsch, who pioneered the universal quantum computer (which you will encounter again in chapter 4).

Universal computers are capable of performing *all* the computations permitted by the laws of physics. Once a universal computer is constructed, all you have to do is to load it with the right programme, and it can simulate any other system that is physically allowed. This includes the biosphere, with all its splendid richness of animals, plants, and microorganisms; and, in principle, it even includes your brain, together with thoughts and emotions.

Now I'd like to let you in on a secret, one that is possible to grasp thanks to the newly introduced idea of interoperability of information. The fact that universal computers are not forbidden in our universe requires the interoperability of information; if interoperability did not hold, computers would behave in a radically different way—and truly universal computers would not exist. It is interesting to see how the counterfactuals in the interoperability of information permit the counterfactuals of universality. I shall give you an example to zoom in on this fascinating fact. Imagine a universe with a 'Dust sector' made of something that we can call (borrowing the terminology from Philip Pullman's trilogy *His Dark Materials*) 'Dust'. Suppose that Dust is a kind of matter that at best interacts *very weakly* with ordinary matter. (Some cosmologies assume that something very close to this imaginary example exists in our universe—which goes under the name of Dark Matter. I would like here to entertain the thought about a hypothetical conjecture without discussing these cosmologies and whether they are true or not!)

Suppose there could be information media made of Dust. This would mean that an information medium in the Non-Dust sector of the universe would *not* be interoperable with an information medium from the Dust sector, because the communication between the two sectors is at best highly imperfect, or perhaps nonexistent. The Dust sector could be just like ours—and have people, civilisations, computers; but they would not be able to signal to us, nor us to them.

This would mean that a bit taken from the Dust sector and a bit taken from the Non-Dust sector would violate interoperability; they would not be interchangeable with one another. In that case, the information embodied in the Dust bit would *not* be copiable onto the other type of bits, and vice versa.

To see the consequences of this conjecture for universal computers, imagine in the Dust sector there was a computer that is universal in that sector. That means that it can perform all computations that are permitted by the laws of physics in that sector, the set D. In the Non-Dust sector, there is another universal computer that can perform the set of all computations permitted in the Non-Dust sector, ND. These two computers have two different repertoires that, together, make up the whole set of possible computations in that universe (D and ND). In our universe, where interoperability holds, two such computers can be put together to create a computer that is universal for the whole set of possible computations. This can be done easily, as I explained earlier: given an input and the specification of what computation one wants to perform, one calls on one or the other computer, which then provides the desired output.

But in the universe with the two separate Dust and Non-Dust sectors, it is impossible! Since the input itself must be encoded either in the Dust or the Non-Dust sector's information media, it can be seen or read out only in that sector and not in the other. Suppose, for instance, the input is encoded in the Dust sector, and the desired computation is in the repertoire of the computer in the Non-Dust sector. As a result, it is impossible to present the input to the computer in the Non-Dust sector; this in turn implies the impossibility of constructing a unique computer that performs all the computations that are permitted in that universe. A universe where the interoperability property is violated would not have universal computers. This fact is a demonstration of how far-reaching the consequences of the interoperability of information can be.

You have traversed several pages in order to understand the connection between physics and information. In what way are you now closer to understanding physical reality? You have discovered, by considering the two counterfactual properties that characterise information media, a key feature of our universe—interoperability, without which what we have been calling 'information', and communication thereafter, would not be possible; nor would computers, let alone universal computers that work in the same way as they do in our universe.

What you have just seen is an example of the explanatory power of interconnected counterfactual properties, this time all related to information. We can think of them as arranged in a pyramid structure. At the base, you find the counterfactual properties of information media: that the flip and the copy operations are both *possible* on some physical systems—information media. On top of this, there is the interoperability of information media: information is *copiable* from any information medium to any other, no matter what type of physical support embodies it. At the very top is universality—the possibility of universal computers. Each counterfactual is needed for the higher-level counterfactual. In turn, these counterfactuals enable a vast number of *other* transformations to be possible. All our information-related technology is based on the interoperability property; so are the most fascinating properties of life and intelligent life—from the possibility of self-reproduction to the possibility of thinking. Remove some of these counterfactuals, and you wipe out all these properties, too.

What's more, by referring to information media and their

counterfactual properties only, without referring to specific ir-relevant details about their embodying systems, we are able to attain a greater degree of abstraction, going deeper than all our existing physical theories. If you remember, at the outset I noted that the elements in the Sentosa landscape are described by very different theories, in the traditional conception. But with the view from counterfactuals, we understand the sense in which some of them are, in fact, very similar: they are all information media. The traditional conception of physics cannot express this fact, whereas the Science of Can and Can't can do so, elegantly and simply.

The approach with counterfactuals also frees information from subjectivity. When we say that some set of states can be cop-ied, we do not need to refer to any conscious subject or observer performing that transformation. A simple chemical reaction where the structure of some crystal is replicated over and over again im-plements the copy operation, and it can do so in the absence of a guiding entity.

The objective, counterfactual properties necessary to explain information are remarkably elementary, and yet they have far-reaching ramifications. Whether you are sitting in a coffee shop drinking coffee while listening to your favourite music, sitting in your armchair scrolling through your phone or reading a book, or watching a beautiful sunset from your balcony, all these phe-nomena can occur because those two operations—the flip and the copy—are *possible*, and because of the interoperability of information media. Both you and I are enjoying the far-reaching power of those counterfactuals right now; I, while writing these lines, and putting a full stop here. You, while reading these very lines, and turning the page to discover what comes next.

Intermezzo Veneziano

"Festina lente, *Francesco. Remember our motto. We must work steadily, but not in haste." Pronouncing those sententious words, Aldo Manuzio sat back comfortably in his leather armchair and cast a long look at his collaborator, Francesco da Bologna, a young punch-cutter. Francesco, sitting in the guest chair of Aldo's study, returned his stare. He looked unconvinced.*

Aldo continued: "Printing and selling books requires careful planning. We are treading a completely new path, and we must be careful not to make mistakes. Just a decade ago no one thought that there could be a trade in books. And now look at how well our press operation is doing here in Venice. We are leading the entire world with our example! Now we need to explore even more adventurous typefaces; we need to expand the market further and make books portable. This may take a little more time, but it will pay off."

Francesco remained silent for a while. He was examining the sketches that Aldo had given him earlier. They showed the letterforms of the new font that Aldo intended to use to print a new kind of book—smaller and perhaps more affordable. Francesco was meant to turn those sketches into metallic punches, each carrying one letter on top. He would have to design elegant, readable letterforms and then reproduce them faithfully in metal. The sketches that Aldo produced were sinuous and elongated, slightly slanted to the right to reproduce calligraphic handwriting.

Francesco finally expressed his fears: "It is not easy to cut such a small shape into the metal, my friend." He was caressing the back of his neck with his right hand, doubtful. "It will be laborious. Making the characters so small and wavy—even smaller than the previous font we designed—is a real challenge. I am worried that it will take too long to come up with a new complete set of punches for the printing machine. We will lag behind with the production. Could we not just stick with the fonts we already used for the other books you printed? The marketplace here moves fast. Other press operations may overtake us while we are trying fancy things like new typefaces."

Aldo sighed. "I know, Francesco—it's difficult to slow down at this point. Venice is very competitive, you are absolutely right; we must not be delayed unless it's for a good reason. You must be feeling as if my ideas are only wishful thinking..."

Francesco said nothing, and Aldo went on.

"*But it is not so. Reproducing the calligraphic style will make a huge difference. I need your skilful hands to design and cut those type blocks properly. Without you, nothing really works here.*"

"*I trust your intuition, Aldo. I have witnessed the remarkable success of your printing enterprise over the past five years. I am just sceptical that this further minute improvement will make things so much better. Is it really worth spending all this time and effort?*"

Aldo looked at him with an odd mixture of rapture and passion. "Yes, Francesco! Yes, I have no hesitation. It must be done. The time is right. With this new calligraphic type, we can produce small books whose content will look almost exactly like that produced in the monasteries by the copyists. Except that they will be available in greater number, for the benefit of many more people. They will be multiplied! Multiplied by our printing machines. We will not just copy one single manuscript into another one like copyists used to. We will simultaneously print dozens of these small portable books; they will be circulating among the public. We will target all the readers of Venice, and beyond. We will inundate them with wonderful stuff. Greek, Latin, vulgar, Hebrew—any language. We could even print grammars and other simple texts in the cheapest editions, not just precious edi-

tions of texts that would be read only by the learned. *Imagine, they will all look beautiful, almost as elegant as if written by hand!"* Aldo paused to catch his breath. *"We are so close, so close! I can almost see all the books that we'll print in this lovely new type. I just need your phenomenal skills—and those of your craftsmen—once more."*

Francesco sighed. He was honoured to be recognised by this powerful man who had ignited the book trade in Venice. Besides, Aldo's words had the ring of truth.

Aldo went on: *"Look—you and I will take an unprecedented step with this font. Once we can publish books with a slanted type-face, which reproduces handwriting, we will have made such an important contribution to humanity. Everyone will use these fonts. And the books will be small. The readers will be able to carry them around with them. It's a fantastic leap. The heart of all this will be our printing machines; the act of copying, in its simplicity; the elegant design of the characters that you and your collaborators will create. You and I will have set the style for the next decades of book printing. Maybe more than decades, I don't know."*

"Very well", the punch-cutter finally said with a smile, *"I will give some thought to how to design the letterforms. I'll bring you samples next time. I'll try to work swiftly."*

After seeing Francesco out of the printing office and closing the main door for the night, Aldo wandered back to his study.

Information

Everyone was resting in the house, other than him. He opened a book that they had printed a few years earlier, containing treatises by Aristotle. It featured Francesco's roman characters. He held the book with a tender and delicate touch, as if it were his lover's hand. He opened it and passed the tips of his fingers slowly over the printed letters, page after page. He felt the patterns of the characters imprinted by his own printing machine. His mind was far, far in the future.

He imagined the future of books, centuries from then. He had more questions than he could answer. What materials will be used to print them? Will the Church allow the printing of impure books, or will there still be hard restrictions? Will the diffusion of books help more people to learn how to read and write? Could one perhaps even make the consultation of books free of charge?

To Aldo, the most beautiful idea of all was that all those amazing possibilities were based on highly precise copying machines. Aldo had been fascinated by the invention of the printing press since the very beginning, when he first learnt about it from the Germans who had brought it to Venice. He had immediately seen the sweeping potential of copying characters mechanically, on such a vast scale. Suddenly so many things had become accessible, once that apparently small and straightforward step had been accomplished. Copying. So simple an action, such far-reaching consequences. Humans had been astonishingly favoured by Nature, Aldo then reasoned, that She allowed the remarkable action of copying to be implemented with such accuracy.

"Maybe, in the future, it will be good to think of a name for the stuff that can be copied", he thought. "I must remember to look for a good name."

The night was gently wrapping Venice in its damp, dark cloak. With a smile lingering on his tired face, Aldo Manuzio slowly slipped into sleep, dreaming of a dazzling future built on knowledge, books, and on his copying machines.

4.

Quantum Information

Where I explain how systems capable of carrying quantum information are information media with two additional counterfactual properties: the **impossibility** of performing certain copy-like transformations and **reversibility**; where you encounter the **universal quantum computer**—a universal computer that harnesses the full power of quantum theory.

S outhwest of my austere and industrious hometown, Torino, along the narrow strip of land where the rocky feet of the Maritime Alps meet the blue, ever-moving immensity of the Ligurian Sea, there stands a round hill, commanding the local bay, and on its top is perched an ancient medieval village called Cervo. My parents and I used to spend our summers there: we lived in a tiny apartment whose centrepiece was the balcony over-looking the beach and a small harbour. From there, you could watch the boats of the fishermen sail about; the seagulls and ospreys dive to catch their prey; the purple and red hues of the everlasting summer sunsets. Sometimes, when the weather was exceptionally dry and the mornings were crystal clear, you could even see the peaks of Corsica emerge in the distance, just over the horizon. They looked like an enchanted fortress surging above the faint dividing line between the sea and the sky.

My summers in Cervo mostly consisted of fun-filled time spent outdoors, in the glittering sun. I used to play with a crowd of local children and with the person who later became my best friend, Francesca. We were chasing the wind and the waves, running around on the rocky beach and into the sea, exploring everything there was to be known, and going beyond that, with imaginative plots involving pirates and treasures lost under the sea (which, alas, we never found!).

When the weather was less good, as a diversion I used to go on walks inland with my mother—who has always had a particular talent for unearthing intriguing finds in the most unexpected places. One thing she discovered there was a farm where an old lady sold eggs, honey, and vegetables. The path to the farm traversed a green, scented forest of maritime pines and then several *uliveti* (olive groves). Turning and looking down, back along the path we had walked, we could see the sea scintillating in the distance; all around, the buzz of thousands of cicadas permeated the air.

It was that grumpy old lady who introduced me, when I was seven or so, to a game of chance. It was a variant of what is usually called the shell game. She would set two cups upside down on a table; then, without me seeing, she would hide a shiny glass marble under one of them. If I could guess correctly where the marble was, I could keep the marble. If not, I had to help her collect the eggs from around the farm. I did not mind that outcome too much; in fact, I quite liked the hens and their chicks. The game was well worth the risk.

The risk in games of chance is due to a counterfactual property. It is the impossibility of correctly predicting something with certainty. I shall call this property *unpredictability*. In the shell game of my childhood, what is unpredictable is the position of the marble relative to the cup. Interestingly, in this case, the unpredictability is not objective: it exists only for the player. From my point of view, the probability of correctly predicting the marble's location was ½. But the old lady, who had the full details, perfectly knew its position. The unpredictability in this game is therefore apparent to the player only because he or she does not have complete information. The person who sets up the game, by contrast, sees an entirely certain, predictable, and deterministic situation.

It seems that, just like in this game with marbles and cups, most unpredictability in everyday life is due to lack of information. When the weather forecast is uncertain, and the weather unpredictable as a result, it is because the information about the initial condition of all the particles in a given region of the atmosphere is imperfectly known. Coin tosses are unpredictable because the initial conditions of the coin and environment are largely unknown. The degree of unpredictability is then quantified with probabilities. The probability of some unpredictable event happening expresses the extent to which one expects it, given what one knows. When asked "What will the weather be like tomorrow?" you can reply, for instance, "I don't know for sure. It's unpredictable. But there is a ninety percent probability that it is going to be sunny." And so on.

All unpredictable behaviours were once supposed to be the same as in this game: not objective, but tied to a specific viewpoint. If given complete information about the actual state of affairs, there is no unpredictability; the latter appears only if one has incomplete information.

Though this belief might seem intuitively true, it was wiped out by the discovery of quantum theory in the first half of the twentieth century. In quantum theory, unpredictability does not arise just from a lack of information; it is inherent to the physical world, even when everyone has all the relevant information. It is objective.

It is rather unfortunate that quantum theory has acquired, in the collective imagination, the status of a quirky beast that is incomprehensible but worthy of attention because of its weird and bizarre demeanour. You may recall the words "spooky action at a distance", used by Einstein to describe quantum entanglement; or the creepy idea of locking a cat inside a box with poison (the notorious thought experiment that Schrödinger envisaged to illustrate quantum superpositions). With catchy phrases, the press has promoted the view that quantum theory is destined to remain a complete mystery. Leaving the good old days of Newtonian physics behind, when the world used to make sense, we now have to resign ourselves to a new and alien picture of physical reality, which accords with the experimental evidence, but whose explanation of the universe is baffling.

None of this is even remotely fair to quantum theory. Yes, the quantum world seems bizarre and counterintuitive at first; but in reality, it is fascinating, subtle, surprising, and, above all, not mysterious—it can be understood. The more it is understood, the more exciting it becomes.

It is true, though, that quantum phenomena cannot be expressed entirely in terms of familiar analogies. As you are about to discover, there is a genuinely new, dazzling set of properties that quantum systems have, which are conceptually far from our everyday worldview. And they are all based on a set of simple counterfactuals, which I shall now explore.

Incidentally, quantum phenomena are important for the progress of our civilisation because they allow for the enhanced information-processing capabilities of quantum systems, which outclass non-quantum information media, such as those deployed by the classical computers we use nowadays. The quantum counterfactuals are the fuel for the next technological revolution, the universal quantum computer. A universal *quantum* computer is a universal computer (i.e., a computer that is capable of performing every computation that is allowed by given laws of physics, as I explained in chapter 3) that relies entirely on quantum theory for its information processing. The computers we currently use are classical because they rely not on quantum phenomena to perform computations, but on entirely classical mechanisms.

The theoretical description of the universal quantum computer has been around since the 1980s, and its features are very promising on paper. It has superior computational abilities compared with classical computers, because its elementary information units (the quantum bits) can explore a much richer set of possibilities than simple classical bits. The effect of this richness, which is entirely due to quantum physics effects, is that the universal quantum computer can be faster and more efficient than a classical computer when it comes to certain computational tasks (searching a large database, factoring a number, etc.). What is important here is for you to have in mind that the leap in possibilities that a universal quantum computer would bring about is analogous to that brought about by the introduction of the classical computers in the first place: it will make an entirely new class of technological improvements possible in the realm of information processing. Alas, the actual realisation of a universal quantum computer has proved to be very challenging in practice. This enterprise has engaged some of the finest minds among physicists, engineers, and material scientists. Numerous IT companies such as Google, IBM, and Microsoft, and start-ups as well, are now trying to race towards the first viable prototype of this machine. But we are still quite a way away, even if we are definitely getting closer.

I belong to the cohort of people who look at technological developments in awe, with optimism and high expectations, but are ultimately more interested in the foundations of the theories that allow such technology to exist. What is it precisely about quantum media that makes them capable of supporting such super-efficient quantum information processing? What can one learn by looking at the foundations if the technology is already pushing ahead?

In fact, by digging into the foundations of quantum theory, we stumble upon a surprising fact: all properties of quantum systems (which are crucial for the universal quantum computer and the related quantum technologies) rest on a few elementary counterfactual features. In chapter 3, I explained that information media are systems with attributes that can be flipped and copied. Quantum systems have these counterfactual properties and therefore are information media in this sense; but they have *more* counterfactual properties, making them so much more powerful.

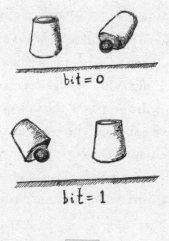

To see what these properties are and understand why quantum systems are a more powerful kind of information medium, I shall invite you to look again at the game of chance with cups, now through the lens of information theory. The two cups together with the marble are, using the terminology of chapter 3, an information medium. They can contain a bit of information, encoded in the position of the marble—as you can see in the figure on page 111. When the marble is under the cup on the right, it encodes the value 0; when it is under the cup on the left, it encodes the value 1.

You can imagine a standard procedure to set up the game: first, toss a coin; if the coin shows heads, put the marble under the cup on the left; if it shows tails, put it under the one on the right. At this point, for the player the bit is perfectly randomised and maximally unpredictable, because it could hold the value 0 or 1 (the marble could be either under the cup on the right or that on the left), each with probability ½. For the person who sets up the game, however, the bit has a definite value (either 0 or 1). They know where the marble is.

Now imagine setting up the game with systems that behave according to quantum theory—not simple information media, but *quantum information media*. What would change in the game?

Our quantum game involves a *photon*, a quantum particle of light, instead of the marble; and two possible paths, instead of the cups. A source emits the photon; then the photon can travel straight along a horizontal path or along a vertical path. The photon and its path constitute an information medium—they can encode a bit: if the photon travels on the horizontal path, it encodes a 0; if it goes on the vertical path, it encodes a 1.

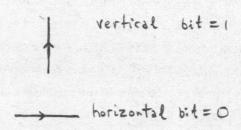

It is possible to set up the game by following the randomisation procedure I explained earlier—but this would not lead us to a quantum game. For example, someone sets the photon to travel vertically or horizontally, according to the output of a coin toss. Again, the chances are ½ for the player to guess correctly which path the photon is on. As you can see, this version of the game is not different from the marble game because it does not use the quantum properties of the photon in any way. We need to explore some other kind of setup, using the quantum properties of the photon.

What is the quantum stuff that a photon can do, while a marble cannot? It can be prepared in states that are exclusively quantum—in the sense that they do not exist according to classical physics but only under quantum physics. An example of these chiefly quantum states, which is relevant for the photon in this example, is what I shall call a *superposition of the horizontal and vertical path*. In order to understand what kind of state this is, how it is related to the states where the photon is on a definite path, and what counterfactuals have to do with all this, we need to look at a definite experiment where the photon is prepared in a superposition of different paths, and then certain measurements are performed on the photon. This will tell us how the superposition is in a certain sense similar to the state of the marble but is fundamentally very different.

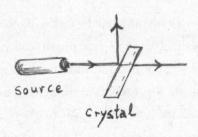

The photon, after having been emitted by a source, can be put into a superposition of paths by guiding it through a special kind of crystal, which, when interacting with the photon, causes it to 'split' along two paths (horizontal and vertical, as in the figure). If you were to guide a beam of light, made of lots of photons, through this crystal, you would see that the beam is split across the two paths (horizontal and vertical)—which is why sometimes this crystal is called a beam splitter. But here I am talking about a single photon, not a beam of light; and what it means for it to be 'split' can be understood only with quantum theory.

So what do we mean by the photon being split across two different paths, in a quantum superposition? Well, one key aspect of the superposition is that after passing through the crystal, the photon could be found with probability ½ on the horizontal path and with probability ½ on the vertical path. When the photon is in a superposition of different paths, it is *impossible to predict* with certainty which path it is on. Its path-location is unpredictable. Experiments happening daily around the globe in the quantum laboratories confirm this behaviour to the highest precision.

Does this mean, then, that when the photon is in a path-superposition, its properties are exhausted by saying with what probability the photon is on one path, and with what probability it is on the other, just like for the marble in our classical example with two cups? No. In reality, the story is much more subtle. Quantum superpositions are not all about probabilities. The photon is not a randomised bit—even if, in some instances, it can look like one.

The first difference from a randomised bit (such as the marble) is that *no one* can predict with certainty which path the photon is on. Even the person who prepared the photon superposition cannot predict that! The unpredictability of the photon path is absolute, unlike in the case of randomisation. The person who prepared the photon knows all the details about the situation yet cannot predict where the photon is. The photon is in a quantum superposition of two different paths; when in that state, it does not have a definite position; and if you were to measure its position, that measurement would have an unpredictable outcome. So, the quantum unpredictability associated with superpositions does not come from the lack of information about the preparation of the photon. If the lady of my earlier story had been using a photon instead of a marble to set up her game, she, too, would not have been able to tell where it was, even though she prepared the photon in the path-superposition herself! This is a striking difference from the marble case.

The second significant difference can be seen by *repeating* the game after having played it once. For the marble, you randomly put the marble under one of the two cups by tossing a coin, as I said earlier; then you repeat the coin toss and, according to the coin value, reposition the marble under one of the cups. The player has still the same chance of guessing correctly where the marble is, ½. Indeed, adding uncertainty to an already uncertain situation can only make things equally, or more, uncertain. Randomising once or twice (or a hundred times) leaves the unpredictability the same. Repeating the steps of the preparation does not change the odds of guessing correctly.

Now look at what happens with the photon. To repeat the preparation of the quantum superposition twice, you need to let the photon through the crystal twice. In a real experiment, this can be done by arranging a second crystal after the first and by setting up a system of mirrors, so that the photon would go through

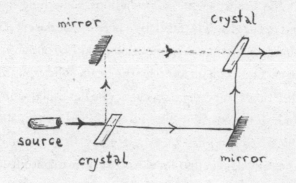

the first crystal, bounce off the mirror, and go through the second crystal, when travelling on *both* the vertical and the horizontal path.

Which path will the photon be on after encountering the *second* crystal? If you think of each crystal as randomising the photon path, judging from the marble example, you would expect the photon to be found with probability ½ on one path and ½ on the other. But this is not how it goes! The photon, after encountering the second crystal, will invariably end up on the *same path* as it was before—the horizontal one. If you believe that the crystal is simply randomising the photon path, the fact we have uncovered seems to imply something impossible: that applying the same randomising procedure (the crystal) twice produces something certain, not random! This conclusion cannot be true. If it were, you could go to a casino and always win simply by waiting for the dice to be rolled twice or the cards to be shuffled twice. We have reached a contradiction.

As always happens with contradictions, something in the assumptions has to give. The revelation here is that the crystal is not a randomising operation, even though it looks like one at first. The quantum superposition created by the beam splitter, unlike coin tosses, dice throws, and the like, cannot be described with probabilities only. It is something else. I first properly learnt about this surprising fact during my doctoral studies, from Artur Ekert— who in the early days of his pioneering work on quantum cryptography had to think hard about how to explain to an incredulous scientific community what was so special about quantum systems, as opposed to simply randomised phenomena. The explanation for this counterintuitive phenomenon resides at the heart of quantum theory—so much so that this fact alone can be taken as a signature of the photon being a genuine quantum-mechanical particle.

What makes the photon so different from the randomised marble is that once it has gone through the *first* crystal, another one of its physical properties (not the position or the path) *is perfectly definite*. The property is that of being in that particular superposition of paths. What's more, there is another counterintuitive fact: letting the photon through the second beam splitter, and then measuring where it is, is one way of measuring that other property (i.e., measuring which superposition the photon is in). That is why we see a definite outcome after the second crystal: a crystal creates a definite path-superposition; then another crystal and a subsequent measurement of where the photon is constitute, jointly, a measurement of which superposition the photon is in! The outcome at the end is definite because after the *first* crystal the photon is in a definite path-superposition (but does not have a definite path).

So there are two properties of the photon that play a role in the experiment with beam splitters: the property "which path" (P) and the property "which path-superposition" (PS). Counterfactuals come in at this important point to explain the relation between these two properties in the case of quantum systems, such as the photon. If it is *possible to predict* the value of P (horizontal or vertical) with certainty, it is impossible to predict which path-superposition the photon is in, and vice versa. When the photon is predictably travelling along a definite path, and P is 'sharp' (in the sense that the photon has a definite value of its position), the other property, PS, is not sharp: a measurement of PS yields an unpredictable outcome. But when the photon has gone through the crystal once, the outcome of a path measurement becomes unpredictable (its position P is not sharp anymore); whereas the other property, PS, has become sharp: the outcome of its measurement is predictable. This relation between the two properties P and PS is based on counterfactuals, and, as I shall now illustrate, it is at the heart of the notorious "Bohr complementarity" displayed by quantum systems: the fact that different properties of quantum systems (such as energy and position, or P and PS) *cannot be simultaneously measured* to arbitrarily high accuracy.

What does all this tell us of the information-theoretic properties of the photon? The counterfactual property I have just uncovered provides the key to answering this question. In chapter 3, copiability emerged as one of the characteristics of information media. Now you are about to discover that the copying task is much more ubiquitous than it may seem at first: it does not pertain to the world of computers and digital machines only. The task of *copying* and that of *measuring anything physical* are fundamentally the same! An apparatus that measures a given property is a system that, when given in input some system, provides in output the value of the relevant property of the system as it really is. A kitchen scale is a familiar example of a measuring apparatus that measures the mass of things. When given some amount of flour, for instance, it gives an indication of the mass of the flour. If it is a perfect scale, when given in input 1 kg of flour, it should give as a reading exactly 1 kg on its display. When provided, say, 10 kg of flour, it should read 10 kg exactly. And so on. The transformation that the scale realises has precisely the same form as a copy operation, because it *copies* the value of the mass given in input into the display of the scale! You have just encountered a new, important fact—a fundamental link between different counterfactuals: *things that can be copied can also be measured, and vice versa.* Another example of a measuring apparatus is the measurer of the photon position. It is a device that when given in input a photon travelling on the vertical or horizontal path, as in our earlier example, can display a message saying, respectively, 'photon on the vertical path' or 'photon on the horizontal path'. Once more, this apparatus copies the value of the path (horizontal or vertical) from the photon onto its display.

Given that 'copiability' and 'measurability' are the same property, I am going to call all properties that *can be copied*, such as P and SP in our example, 'observables', hinting at the fact that they are things that can be measured (i.e., observed).

Back to the two physical properties of the photon: the path observable (P), and the other observable (the superposition of different paths, PS) can encode one bit of information each, because each observable P and PS can each be copied and measured individually. Earlier I said that P and SP cannot be simultaneously measured to arbitrarily high accuracy; or, equivalently, their measurements cannot both have predictable outcomes: if you *can* predict the value of P (horizontal or vertical) with certainty, you *cannot* predict which value PS has with certainty, and vice versa. Given the fact that copying and measuring are equivalent, this fact also means that P and SP are *not simultaneously copiable*. So, thanks to our information theory based on counterfactuals, we see now that the *impossibility of copying P and PS simultaneously* implies that they cannot be used jointly to store information simultaneously to arbitrary accuracy. Another way to think about this is that if you have a measurer, or a copier, that works well for P, then it must give a more imperfect, less accurate reading or copying of the other property, SP, and vice versa. So P and PS cannot be used jointly to signal—if you can signal reliably with P, you will not be able to use the same system to signal reliably with SP, and vice versa.

This fact does not apply just to photons and their properties. It applies to any two properties of any quantum system that are not just deducible from a single property. For instance, the velocity and the position of an electron around the nucleus of an atom are both observables, but they cannot be simultaneously measured to the same accuracy. When the velocity of the electron can be measured accurately, its position cannot be. In the early days of quantum theory, this fact gained the name 'Heisenberg uncertainty principle'. Later, it became known as a special case of the 'no-cloning' theorem for general quantum observables. It is a pillar of quantum information; it distinguishes quantum from non-quantum systems.

So, here is an important and fascinating conclusion: quantum systems have at least two observables, such as P and SP, which are *not copiable jointly*, or not measurable jointly, to the same arbitrarily high accuracy. It is a counterfactual property, to do with what *is impossible* to perform on quantum systems. It also constitutes the crucial difference between the classical unpredictability of a coin toss and the quantum unpredictability arising with quantum superpositions, as for the photon path.

Is that all there is to quantum systems and their counterfactual properties? No, there's more. You need another counterfactual property to capture quantum information media—this time about possibility. You need *reversibility*.

Reversibility in physics usually refers to the *possibility* of reversing a transformation. A transformation is physically reversible if, whenever there is a way to perform it, performing it in the reverse direction is also possible. When you cross a bridge from one side to the other, you are performing a reversible transformation (assuming the bridge does not subsequently get destroyed). Flipping a bit from 0 to 1 is also reversible. But cooking an egg is not a reversible transformation; nor is splatting the egg on the floor.

A photon, when it behaves in a quantum way, must have the counterfactual property that *all transformations allowed on it can be reversed*. So if you apply, for example, the crystal on the photon, you should be able to use the crystal in reverse.

So here is the main conclusion of this discussion. A quantum information medium is a system with the following counterfactual properties:

1. It has at least two information variables (such as P and PS) that are *impossible to copy simultaneously* to arbitrarily high accuracy (non-copiability of information variables).

2. It must be *possible to reverse* all the transformations involving these variables (reversibility).

The smallest unit of quantum information is a quantum bit, a *qubit*. Photons, electrons, and other elementary particles can all be used as qubits. The reason why perfect qubits are hard to come by in everyday life, and the universal quantum computer is very hard to realise in practice, is that accurate reversibility is extremely difficult to achieve in practice while at the same time preserving the other quantum properties of the physical object in question.

Quantum theory says that is *possible*, but it arises only under carefully controlled circumstances. Most photons around us, such as those emitted from the sunlight, do not undergo reversible transformations when left to naturally occurring conditions. They would if they operated under perfect isolation; but usually they interact with many other particles, in an uncontrolled fashion (another manifestation of no-design laws, as seen in chapter 1); and all such interactions should also be reversed, if one wants to achieve perfect reversibility of the photon. It is technologically very challenging to do that: there are far too many interactions to easily keep track of them all! As a rule of thumb, the larger the system in question, the harder it is to achieve reversibility while still keeping the other quantum properties. For other elementary particles, those with a mass, it becomes harder than for photons, but it is still achievable under highly controlled conditions in the laboratory. For molecules that are as large as a virus, for example, reversibility is, at the time of writing, just outside our technological reach. The reason is that all the subparts of large molecules have multiple opportunities to interact with other elements of the environment, and each of these interactions makes the reversibility harder to achieve, because in order to reverse the transformation on the particle in question, one also has to reverse all the other tiny interactions with the environment. By interacting with the environment, the particle has exchanged information with several other entities; and unless *all* those information exchanges are also undone, the particle cannot go back to the original state.

Given that there are gazillions of possible encounters with the surroundings in a single quantum experiment, one faces a formidable degree of complexity, even with relatively small systems. Right now, my colleague Markus Arndt and his group in Vienna are developing wondrous machines capable of persuading particles to achieve this reversibility while maintaining their delicate quantum properties. At Google, IBM, and Microsoft headquarters, tech engineers are trying to achieve this reversibility on a large scale, in order to produce ever more reliable quantum computers that exploit quantum phenomena.

If one can achieve *both* the reversibility and the impossibility of joint copying of the information variables of a quantum system, then a number of other surprising quantum properties also follow, which are a characteristic sign of the ability to process quantum information. The most prominent one is *entanglement*.

Entanglement is one of the most exotic and powerful—and misunderstood—properties of quantum systems. When its properties were fully discovered, it soon became clear that it would revolutionise the way we understand composite quantum systems—that is to say, systems made of two or more subparts. The concept was already known to the pioneers of quantum theory—Schrödinger is usually credited with introducing the idea. But the full potential of entanglement-based ideas was unleashed in the early days of quantum computing, when that was first considered as a *resource* for quantum computations. The physicist Vlatko Vedral, who in the 1990s pioneered our most subtle measures of entanglement, often jokes about the explosive development of the field, remarking that he managed to get an academic job simply by working on the foundations of this elusive and fascinating quantum phenomenon. (Those were times when working on fundamental and adventurous topics was still encouraged in academia; in the current academic monoculture, pursuing transformative and risky projects is, sadly, becoming harder and harder.)

Entanglement arises when you have two or more quantum entities interacting; for example, two photons; or an electron and a photon. The essential feature of entangled quantum systems is that the information one can gain by jointly observing the systems is *more* than the information obtained by observing each system separately. This phenomenon is rather counterintuitive and removed from everyday experience. If you consider, for instance, two cells on a Go board, each can be in one of two states: empty or full. Each of them is a bit. If you put the two cells together, there are now FOUR possible states they can be in: both empty; both full; the first empty, the second full; and the first full, the second empty. These are all distinguishable from one another by just looking at the state of each individual cell. Now, if you replace each cell with a qubit—a photon, for instance—something *more is possible* when you consider the two qubits together: there are more states available for those two qubits than just the four states you can construct by considering their information-carrying attributes separately. The additional states correspond to the situation where the two qubits are entangled. The characteristic property of these states is that they can be used to encode information, just like, say, the four states of the two Go cells; but unlike the Go cells, the states cannot be distinguished just by looking at the state of each of the two qubits separately. So, when the qubits are entangled, it is *possible* to extract information *globally* (acting on both qubits) but *impossible* to do so *locally* (acting on each qubit separately). This fascinating fact is not just curious, it is actually practically useful—for instance, if you wish to hide information in the two qubits, like in a safe. It is the base for entanglement-based quantum cryptography, where entanglement is used in order to transmit information securely, by exploiting the fact that by looking only at the two qubits individually, it is not possible to guess the joint state of the two qubits.

I don't want to discuss the properties of entanglement further; the point I want to make is a simpler one. The point is that the *possibility* of these additional states where two qubits are entangled, which have no classical analogue, stems from the counterfactuals I identified earlier. So here is a fascinating fact: by *reducing* the possibilities on each system (by restricting the ability to copy certain properties, like P and PS, as indicated earlier), you get *more* possibilities when you consider two qubits jointly—the possibility of entanglement. The Heisenberg uncertainty principle—which says that measuring simultaneously incompatible observables is impossible, such as position and momentum, or P and PS in my earlier example—gives to a quantum particle a much wider range of possible behaviours. Declaring something impossible leads to more things being possible; bizarre as it may seem, it is commonplace in quantum physics.

I have singled out the elementary counterfactuals that underpin all the characteristic properties of quantum systems; and I have mentioned how these counterfactuals have far-reaching implications for the possibilities of quantum systems, by, for instance, enabling entanglement and quantum computation. Why is it so important to pin down the counterfactuals at the foundations of quantum phenomena? To explain, we have to enter a terrain which is yet to be fully explored. Sooner or later, quantum theory will have to be abandoned, in its current form, and upgraded to something more accurate and universal. Physicists expect this because quantum theory is incompatible with the current best theory of space, time, and gravity—namely, Einstein's theory of general relativity. The latter accommodates only non-quantum information media within it; it is not compatible with qubits. Gravity as described by general relativity is therefore incompatible with the structure of quantum theory as we know it. Consequently, it is reasonable to expect that both quantum theory and general relativity will ultimately be modified and merged into a more general explanation. Theoretical physics has already achieved a few great unifications: Maxwell's theory of light and Newton's classical physics were incompatible with each other; ultimately, these incompatibilities were resolved by the discovery of quantum theory and general relativity, which merged both of them into a more robust and general explanation of the universe.

With the physics of counterfactuals, features of quantum and classical information media can be expressed independently of the details of quantum theory or classical physics. The copiability of specific properties, the impossibility of copying others, and reversibility are general features about which we can talk without committing to quantum theory explicitly. They provide profound connections between intuitively very distant patches of the fabric of reality—such as photons, electron spins, neutrons, and other particles—that would otherwise look very different, and yet they all have the same set of counterfactual properties. The power of the Science of Can and Can't is that it expresses the essence of quantum systems without ever committing to quantum theory's full machinery (made of specific laws of motion) as a whole. This is important in view of the fact that, as I mentioned, quantum theory soon may be superseded by a better theory. My bet is that even if quantum theory eventually gets wiped away, the counterfactual, information-theoretic structure we have explored in this chapter will remain, because it has deeper foundations than quantum theory itself. These are the features that will survive the next revolution in physics.

At a glance, all around you, things appear superficially very diverse. But when looking for long enough and with the right spirit of scientific discovery, while also asking good questions, and trying to play around with some bits here and there, sometimes we find a shiny, far-reaching connection between things that seem diverse and unrelated; and this connection is based on a unifying explanation of the distant phenomena in question. For example, physics tells us that a specific relation exists between mass and energy; between the finiteness of the speed of light and the structure of spacetime; and (as you have learnt) between measurers and copiers. We have achieved yet another unification with the Science of Can and Can't. Quantum and classical information turn out to be two aspects of the *same* set of information-theoretic properties. Quantum information media are a special case of classical information media, with two additional properties: reversibility and the non-copiability of certain sets of states. Quantum and classical media are different, but perfectly compatible with each other. The fracture between the quantum and the classical world—the former supposedly aloof and incomprehensible, the latter more friendly and intuitive—has been healed.

Realising a unification of this kind goes together with abstracting away irrelevant details, making our understanding more general and robust than it was previously. These are promising features for a deeper understanding of nature, of which you and I are exploring the most basic building blocks. Once the edifice is built, it will be beautiful in its elegance and simplicity, and counterfactuals will be the robust elements of its foundations.

A Flight with the Luckdragon

Anne longed for her English teacher to be interesting, passionate, loving, and understanding. That she was none of the above was something that Anne had learnt to accept over the first five years of school.

The coldness and ill temper of Miss Thornhill—that was her name—had been evident to Anne ever since that middle-aged lady first made her appearance at the school on a sunny, crisp September day. It was Anne's first day of school and she was filled with hopes for the future, especially regarding her English teacher. Anne loved hearing and reading stories, and she had formed the idea that English classes would be all about stories. Anne's reverie collided miserably with reality the moment the teacher stepped into the *classroom. Miss Thornhill introduced herself with a stiff face and tight lips. Not even the hint of a smile. Her gaze never rested on any of the students' eyes. She kept her eyes fixed on a distant point, lost somewhere in the light blue paint of the classroom's back wall. During the first introduction, she launched into a rant against invention and imagination: she wanted her students to forget about all that fairy-tale nonsense, and rather to make sure they knew their grammar properly and were able to tick all the boxes in a test correctly.*

Miss Thornhill's lessons were dry, constellated by notions to memorise. They lacked passion or any kind of intellectual fun. Since that very first day, Anne had known that it would be tough. She wanted to enjoy her lessons, but that seemed to be impossible. She wanted to be able to love her teacher's methods. But that was impossible, too.

Anne was an imaginative girl. Most children are; but Anne's imagination was something far out of the ordinary. Ever since she could remember, she had had stories blooming in her head—and not just the stories she read in books. One of her secret delights was to invent and tell stories to herself, whenever she was alone. She would even act them out, playing the speaking parts of all the characters. Of course, she knew that they were not real; but she greatly enjoyed the practice of the mise-en-scène. Over the years, as she mastered her writing, she started writing down some of the best stories in a secret notebook and playacting some with her friends.

Quantum Information

This was how Anne devised an ingenious strategy to put up with her insupportable English lessons. She came up with a fantasy, to transform what was there in reality into something nicer to deal with. She invented a teacher—the sort of teacher she would have liked to have, and who, like all good teachers, was also a friend.

Anne's imaginary teacher was no ordinary teacher. His name was Falkor the Luckdragon. The idea for the Luckdragon had come to Anne while reading about the homonymous character in Michael Ende's enchanting novel The Neverending Story. *But she gave the Luckdragon a much deeper personality. As in the novel, the Luckdragon was a majestic white dragon, with a splendid white mane and delicate long paws ending in curved, sharp nails. His soul was made of pure thought, fun, and delight. But Anne's Luckdragon had also a deep knowledge of English, science, and philosophy, particularly epistemology. He called himself a Natural Philosopher. Anne heard that term in a history class. The teacher had said that Newton was a Natural Philosopher. Fascinated by that concept, Anne had immediately weaved it into her reverie. So whenever Miss Thornhill did something particularly off-putting, Anne would counteract it by writing a story where things went completely differently. And the Luckdragon would always feature in those stories.*

One particular day, Anne returned home thinking that a story was called for. Miss Thornhill had introduced a set of rules to make the procedure of leaving school smoother (in her twisted mind, that is). Children had to line up behind her in an orderly 'crocodile' (a queue, two by two) and follow her to the gate in total silence. Only once they were at the gate could they break the lines and finally step out of school in their own natural, disordered ways. Even worse, the children weren't even free to choose who to pair up with; that choice was—of course—Miss Thornhill's.

By the time she reached home, Anne had thought of a perfect idea for the story. It was time to invoke the Luckdragon; she also thought of weaving in some mysterious ideas a friend of her mother's had once told her about quantum theory. He was a physics professor, working on something called quantum information. What he had told her was all about quantum algorithms that could run on a futuristic machine called a 'quantum computer'. At the time, Anne had gathered some interesting facts about its inner workings, which she now planned to use in her new story. She sat in her favourite corner of the house with a cup of Earl Grey tea, some biscuits, and milk, and started writing.

I do not remember exactly how the Luckdragon and I ended up talking about that. The topic of the conversation was 'constraints', and it was a late summer afternoon. We were sitting in the small garden behind the Luckdragon's complicated house. The Luckdragon was sipping English Breakfast tea and munching on a peanut snack that I have never seen elsewhere. I, on the other hand, was complaining a lot. I was complaining about constraints.

"You see—I think that adding constraints always makes things worse. I hate it when my teacher comes up with pointless rules that make our life more annoying. Tick this box, tick that other box. Do this, don't do that. What good can there be in all these 'can and can't' that we have to stick to?"

"Well, Anne, I understand the frustration. School rules are often nonsensical. It would be far better that a child could stipulate with their teachers what's best for a particular situation, on an individual basis. But then it would not really be a school anymore."

He sighed gently, slowly scratching his flank with one of his paws, then resumed his thoughts with a laconic statement: "Nonetheless, some constraints are good. Some are wonderful."

I was puzzled. I was not expecting this viewpoint from the Luckdragon—he was all about flexibility and having choices. I thought for a little while, then asked: "Do you mean constraints in art or literature? Yes, OK, but those you can change at will, and you can make them evolve according to what you think is better. So they are not really fixed constraints."

"No—I meant actual, immutable constraints that add to your possibilities."

"Really? That's impossible. How could that be?" I was rather skeptical, but eager to hear more.

The Luckdragon shifted to a more comfortable position while turning his tail in a tighter spiral. His white mane glittered in the late-evening summer sunshine. Meanwhile, the industrious bees were still going about their business, buzzing intensively around a lavender bush.

"Yes," the Luckdragon continued, "I mentioned to you quantum theory—it is one of the most fundamental explanations of the universe currently in our possession. The main requirement of that theory is a constraint. One of the main constraints is that there are states of a system you cannot distinguish perfectly from one another, even though they are physically different."

"You cannot distinguish them perfectly . . ." I repeated slowly to myself. "Does that mean that it is impossible for me to see the difference between them, no matter how hard I try?"

"Yes, it is impossible to perform a perfectly reliable distinguishing process, or a perfect copy-like process, as physicists would put it. So, for example, you cannot use them to signal to a distant boat, say. It would be a very bad idea. You'd likely fail on most occasions!"

"OK. But this is just reinforcing my original point: that constraint is limiting your possibilities—and that is bad, I guess."

"It need not be. It can be the beginning of something remarkable. What happens is that if you take two of the systems with that property together, because of that constraint that you *cannot* distinguish states perfectly, you can create *new* properties. One of them is called 'locally inaccessible information'. That means that you can hide secret information in those two systems in a perfectly secure way. And I mean perfectly. Nothing in the universe can possibly read it, unless you give them the key. This is what quantum cryptography is all about."

"Really?"

"Yes."

I was not quite convinced. It sounded implausible. "How do we know about this?"

"Well, quantum theory tells us that it is like that. In other words, it tells us that locally inaccessible information is possible, and how to create it—it involves a quantum phenomenon called 'entanglement'. And that is all a consequence of constraints about what cannot happen in physical systems—in this case the impossibility of copying."

I was excited. The Luckdragon had just completely changed my mind about constraints.

After writing this story, Anne felt considerably better. She was even able to finish her English homework, despite Miss Thornhill's popping up from time to time in her head. The next day, as she approached the front entrance across the black-and-white tiles of the schoolyard, she stopped. A game had just occurred to her—she began jumping from one white tile to the next, avoiding the black ones altogether. She was imagining that stepping on the black tiles would be like stepping into a bottomless hole. It was fun.

She smiled, as she had found another way in which constraints are good, just like in quantum theory. The impossibility of stepping into the dark tiles was creating more possibilities for fun. Her mind was on fire, flying high in the skies, traversing the clouds at Luckdragon speed.

5.

Knowledge

Where I explain that any transformation that occurs reliably requires
a generalised **catalyst** an entity that is capable of performing the
transformation and retains the ability to cause it again; that any
catalyst must contain an **abstract catalyst**, which consists of
knowledge (information capable of self-preservation).

While exploring the garden behind my grandparents'
house as a child, I made what at the time I considered a
phenomenal discovery. Hidden in the grass there was
a round, almost perfectly circular hole, approximately the size of
a small coin. I could just about glimpse the soil inside it. The hole
sloped down into the earth; it was the beginning of a dark tunnel.
I was intrigued: *Alice's Adventures in Wonderland* started just like
that, with a tunnel in a garden. Unfortunately, my tunnel was too
narrow to lead me anywhere. It looked rather like the house of a

minuscule creature. I gathered some grass and a few bits of leaves in front of the entrance, in the hope of tempting the creature out. Nothing happened, though, so my attention shifted, and I forgot about the hole for that day. Still, the day after, I checked it again. The hole was back in its original state, as round and clean as it was the day before: no grass or debris blocking the entrance, all in order. I was excited. Something was happening, after all.

I repeated the experiment a few times and, invariably, the hole would be cleared in a few hours. The cleaning would happen reliably: the hole was kept the same size, the same shape, day after day. Even after a week during which it rained copiously, the hole was still unchanged. Eventually I discovered that there was something inside the hole, albeit less exciting than Alice's new world. The hole was, in fact, the den of a mole cricket. A mole cricket is not a hybrid creature from some myth, as the name might suggest. It is an insect. Chunkier than an ordinary cricket, it has the appearance of a tiny black armoured vehicle. It digs small holes in the soil, in which it lays eggs, stores food, and even traps other insects by luring them close to the hole entrance. The tunnel is slightly sloping, so that the victims get caught and slide to the bottom of it, where the mole cricket devours them.

I kept trying to modify the hole, until one day the cricket stopped its activity. It left, I conjectured, possibly being annoyed by all that extra work of cleaning the hole's entrance. I knew that it was no longer there, because I saw that my changes to the hole, and all the other changes that were naturally occurring, were no longer corrected. The hole faded away in a few days, under the action of April's abundant rain.

This story about gardens and insects offers a particularly spotless example of a general, counterintuitive fact. Most changes, or transformations, that happen reliably around us *require something to stay unchanged*. In the case of my story, the things that undergo a transformation are the soil and the grass around where the hole is built. The thing that stays unchanged is the cricket. To be precise, the cricket stays unchanged only in a particular respect. What stays unchanged are all the features that make it *capable* of building the hole—for instance, its strong forelimbs, which are shaped like shovels and even armed with sawtooth edges to excavate more efficiently. This guarantees that the hole can be kept roughly in the same shape for much longer than the time over which the environment operates with its inexorable erosion.

As I said, it is a general fact. Any transformation happening reliably with a high accuracy (such as making a perfectly round tunnel) requires something (such as the cricket) that stays unchanged in its being capable of causing it again. The entity must retain that property because that is necessary for the transformation to happen reliably. To clarify the point, in the hole-making transformation, things like enough space and soil, and the cricket, are all necessary. But as I said, it is only the cricket that remains the same before and after the transformation, in some respect: the soil gets abraded away, the space is used up by the hole, and so on; but the cricket, for its life span, is very nearly unchanged in that key ability.

Things that can perform a transformation and retain the property of doing so repeatedly, such as the cricket, deserve a unifying name. *Catalysts* is quite a good one. Here one has to be careful, because I am borrowing the term from chemistry to indicate a much more general class of systems than chemical catalysts. Chemical catalysts are entities that make a chemical reaction happen more reliably and faster, when they are present, and that retain the property to do so over and over again. They operate a bit like facilitators. When all the other reagents are around, but the catalysts are not there, it takes ages for the reaction to happen, and other reactions may well consume the reagents first. But if catalysts are present, then the reaction happens quickly and deterministically. Since chemical catalysts are distinguished from reagents because they do not change, while everything else does, I shall use 'catalyst' to indicate systems that can cause a transformation and retain the property to do so, like the cricket in my story.

Why are catalysts, in this generalised sense, interesting? Answering this question will require one to understand the link between catalyst and knowledge, as introduced in chapter 1.

The reason is that most systems undergo changes during processes that involve them (i.e., they do not stay the same), unlike catalysts. Also, most transformations in physics do not happen reliably. Those that do require a catalyst that can perform them.

Catalysts, and likewise highly accurately performed transformations, are *hard to come by*.

Some of you might be suspicious of the notion of 'hard to come by'. It sounds too subjective. What looks hard to realise for some might be very easy for others. Actually, that is not the case. 'Hard to come by' has an objective meaning, established by the laws of physics. One thing is harder to come by than another, in this sense, if the former requires, compared with the latter, more of what is naturally given by the laws of physics for it to emerge. This notion is objective because what the elementary elements are (things like energy, time, elementary chemicals, and so on) is set by the laws of physics.

Take a look around and see what is not hard to come by. What the laws of physics give us in abundance are things like elementary particles and fields—entities which, as we have learnt via modern physics, are used to explain the existence of elementary particles themselves and their mutual interactions. For example, if we see electrons and protons being attracted by what, classically, we would call the electrostatic force, we do not have to invoke anything else but the laws of physics to explain that attraction. I call things that are given in abundance in our universe, such as fields and particles, 'naturally occurring systems'. And likewise, the interactions that need no more than the laws of physics (as we know them) to be explained fully, I call 'naturally occurring interactions'.

Among these naturally occurring systems and interactions, there are a few accurately and reliably performed transformations.

For example, the planetary orbits around the sun are almost elliptical—so, one could say, the task of making the planets describe elliptical orbits is well approximated in nature. This fact is a direct consequence of naturally occurring interactions, because there is a symmetry in the gravitational potential that causes the orbits to have (approximately) that shape. In the case of such transformations, what has to stay unchanged for them to be performed to a high degree of reliability—the catalysts—are just the underlying physical laws, which do not require further explanation.

But most of these kinds of transformations are not like that—there's more to them; some non-elementary catalyst must be present for them to be performed to high accuracy. Think again of the hole in the grass being formed and reformed. That is not directly explainable given the laws of physics, because there are no naturally occurring interactions that cause tiny holes in the grass to materialise and be maintained to a high accuracy—unlike planetary orbits. The mole cricket's holes in the grass of course obey the laws of physics. But to explain their persistence, we need some additional bit of explanation (involving the cricket). Likewise, to explain most transformations that happen accurately and reliably, we need some additional explanation. This explanation will involve the concept of a catalyst, but also of information, as defined in chapter 3—in terms of counterfactuals. It will involve a particular type of information, which can enable its self-perpetuation (in chapter 1, I called it knowledge). To explain how most transformations happen to a high degree of accuracy, in other words, we need to resort to a new class of counterfactuals.

The best way to define this type of information is that it is exactly the thing one would ultimately have to eliminate in order to prevent a particular transformation from being performed reliably. There are, as I shall explain, two main types of steps one can take to that end. To see what they are, let's return to the mole cricket's den.

When the hole is formed out of a volume of moist soil in the garden, what happens is that a certain volume of soil is converted into a hole and some waste products, such as the soil that is removed to form the hole. For the transformation to happen reliably, one has then to make sure it can happen again, a number of times in succession. This reliability is ensured by the presence of the cricket, which, as I said, is another entity that is necessary for the transformation to happen, but has, in addition, the property that it remains unchanged in its ability to cause the transformation multiple times.

One sure way of stopping the transformation is to remove the available soil: in other words, one can restrict the available materials that are necessary for the transformation to take place. The other way is to somehow modify the cricket, the catalyst, and thus prevent it from doing its task. It would seem that killing the cricket would be enough. But as all gardeners know, this is not quite true. The task of stopping mole crickets from digging holes is much harder than killing a single mole cricket. The cricket will have offspring and other relatives, which can spread and keep digging holes in your garden. And even if you were to destroy all the crickets in your garden, there would be something that still survives—and that is the genome that codes for the mole cricket. The genome is information, as I said in chapter 3, because it can be copied (during DNA replication). So what you really should destroy is that piece of information. If that is not destroyed, the cricket could always be brought back to life via some laboratory experiment, along the lines of what happened in the film *Jurassic Park*, and that would cause the hole-making transformation to happen again and again, as before. So that piece of information contained in the genome is the thing that you would ultimately have to destroy in order for the mole-cricket activity to cease forever.

So the genome is the thing one would have to exterminate in order to stop the transformations caused by a living entity. This fact is encapsulated rather neatly by a curious story. The story was written in an old book of science fiction that my father particularly enjoyed. The title is lost now, but I remember that story with great clarity. At that time, I enjoyed sitting by my father's armchair and trying to cast an eye on what he was reading, eventually inducing him to read out loud. He also used to modify details of the stories as he was reading them, so I am not sure to what extent the story I remember was truly in the book, or was something he invented.

Anyway, the story is about some alien civilisation afflicted by a terrible disease that is caused by a bacterial infection (we can sympathise with them, given our recent experience with the COVID-19 pandemic). They are trying hard to eradicate it, by deploying chemicals of various sorts, but they do not know yet about vaccinations and bacteria and DNA—so they have no idea of how the disease propagates. At some point in the story, a human expedition to that planet shows up, and they decide to help the aliens. The humans' secret weapon is that they know exactly the DNA sequence that codes for the bacterium. Using futuristic technology (which actually we do not have to this day!), the humans programme an army of nanorobots, each loaded with the exact sequence coding for the bacterium's genome. The robots target that particular sequence and destroy it. All the bacteria are exterminated, and everything seems fine. When the human expedition goes back home, some terrorists get into the possession of the programme that was fed into the robots. They use the programme, containing the information about the bacterium's DNA, to produce a variant of it that is resistant to its DNA being manipulated. The rest of the story is about how the humans manage to prevent the spread of the bacterium—there is a happy ending, mitigated, however, by the creepy feeling that the bacterium's DNA is still lurking somewhere, and there might be a pandemic sometime soon.

This story presents another case that is relevant to defining the particular type of information we are interested in. What is that entity that the robots are programmed to destroy, exactly? One could say it is a molecule of DNA, which enables a bacterium to exist and to function properly. This is only part of the answer, because, in fact, the same thing is also contained in the robots' programme—which is not a DNA molecule. It is a string of bits, with a particular set of properties. It is, again, a piece of information with special properties. What are these properties?

First, there is the property of enabling the organism to be formed and to enact all its relevant behaviours. So the piece of information can enable transformations and retain all the relevant properties after having performed the transformations once. Hence, in the terminology we have just introduced, it is a catalyst.

However, the genome is not an ordinary catalyst. It has two additional properties: it is information, because it can be copied in the process of DNA replication, *and* it causes itself to remain instantiated in physical systems over generations because it guarantees that the organism for which it codes can survive in a certain environment. I shall call this type of information an *abstract catalyst*. 'Catalyst', as I said, because it *can enable* transformations and retain the property of causing them again. 'Abstract', because its identity does not depend on the physical systems in which it is embodied—it *can be copied* from one embodiment to another, without changing its properties: it could be in DNA or in a nanorobot's computer. According to our criteria in chapter 3, an abstract catalyst is made of information, for it is copiable. Also, it is information that *is capable of enabling its own preservation*: in the terminology of chapter 1, it contains knowledge.

Before moving on, I want to clarify something important. I just said that a catalyst can enable, or cause, transformations on physical systems. Truth be told, the term *causation* has acquired, especially in physics circles, a bad reputation. Saying that a catalyst has the ability to *cause* certain transformations could therefore be misunderstood for one of these bad ways of looking at a cause. But it isn't. When we say that the catalyst 'causes' a transformation, we mean simply that the transformation occurs only when the catalyst is available, and that the catalyst retains the property of making the transformation occur over and over again.

Although other notions of causes are problematic and seem arbitrary in physics, this one is not because it is clear when some system is or is not a catalyst. For instance, one can say that the catalyst that produced a blue-green algae cell is the parent cell that originated it via self-reproduction. Indeed, the parent cell is the particular cell that is capable of constructing the daughter cell by using the information in the DNA. The DNA and the rest of the cell, in this context, are the only systems necessary to the transformation, which also *stay the same* in the ability to enable the transformation again, before and after the transformation. The retention of this ability is a distinctive and objective feature of the catalyst, which does not apply to other elements in the environment. For any transformation that occurs in physical reality, one can unambiguously identify a catalyst that is capable of realising it and of retaining the ability to cause it again. It is the catalyst for the transformation; one can think of it as a cause of the transformation.

How do we distinguish abstract catalysts from other kinds of information? We need to look for information that can enable transformations and is resilient. Again, biology seems to provide a useful example where abstract catalysts are distinguished from generic information. Think of a plant—for instance, a maritime pine tree standing on the Ligurian coast. Even before visualising the tree, one can perceive its scent. (After the rain, molecules of particular chemicals, called terpenes, are released by the trees and cause the air to be permeated with that characteristic fragrance.) Zooming in, one can take a closer look at the minute green needles that line the pine tree's branches. Going in closer still, one reaches the level of a single cell of the pine tree. Closer still, and one is looking at a DNA strand inside the cell.

Let's look for the abstract catalysts in the cell. Every part of the DNA strand in the cell contains information, in the sense I explained earlier in this book. This is because the strand is *copiable*. It is copied in the process of DNA replication, when cells self-reproduce. However, only some pieces of that DNA strand code for something. These bits are what biologists would call adaptations. An adaptation (as already mentioned in chapter 1) is a piece of information in the DNA with the ability to enable a certain trait to emerge in the organism that hosts that DNA: we say that it codes for that particular trait. For instance, there is a bit of DNA that codes for the colour of the pine needles—such as that particular shade of green that pine trees have. However, not all adaptations are resilient—which is the other salient property of abstract catalysts. To be resilient in a given environment, an adaptation needs to be *useful*: it must increase the probability that the genes that code for it will be passed on to the next generation and preserved for generations, in that given environment. For example, an adaptation that makes the pine tree's needles yellow might not be useful if the environment contains insects that target yellow plants as food; so, although it can enable transformations on physical systems, it would not be resilient, and it would not qualify as an abstract catalyst. Conversely, an adaptation that makes the needles darker might be a useful one in an environment where there is strong sunlight, because it can protect the needles from damaging UV radiation. Being a useful adaptation guarantees the survival of that piece of information with causal abilities—it is what guarantees that it is resilient, and that it qualifies as an abstract catalyst. So the information in a piece of DNA may or may not be an abstract catalyst, depending on whether or not it can propagate itself for generations, thus remaining instantiated in physical systems. Generalising from this biology example, information that can enable transformations on physical systems must also be resilient in order to qualify as an abstract catalyst.

So catalysts are systems that can enable transformations and retain the ability to cause them again. Abstract catalysts are catalysts that are copiable and can perpetuate themselves. They are catalysts made of resilient information, which we call knowledge. Now, an intriguing fact that I am about to explain is that *all catalysts must contain an abstract catalyst*. Catalysts, in other words, must have some properties that are invariant, no matter what transformation they are intended for, and that invariant part must be made of knowledge. As I am about to illustrate, this remarkable fact is due to the particular structure of the laws of physics in our universe.

Let's consider the example of assembling an aircraft from elementary components, in the Airbus factory. The elementary components are the subparts of the plane—such as wings, engines, seats, and wheels—but to be more expansive, we can think of the whole process that produces an aircraft out of even more elementary entities—such as metals, plastic, and similar materials. The whole factory is the catalyst for this transformation.

Where is the abstract catalyst? As I said, it is the thing that ultimately one has to eliminate for the factory to stop working properly. Imagine a slight modification to the factory—for example, by introducing a flaw in its production line. If it is a well-run factory, there is a way to fix the problem and thus restore the production process. So that change is inconsequential for the output. However, if you destroy the sequence of instructions for constructing the plane, or the instructions for repairing the factory, the factory might have to shut down. The recipe for the aircraft must be copied for the factory to survive. It is the abstract catalyst that keeps the factory going for years.

This recipe is the set of instructions to realise the construction of the aircraft (to the accuracy set by the factory's standards). It is a recipe in the sense that it is a sequence of steps that one has to follow in order to forge the metals into the shape of a plane. The recipe for a fully fledged aircraft is what allows the construction of the aircraft to happen reliably, because the final product is checked against the procedure until it meets the criteria set by the quality control of the company. Also, the recipe is preserved down the line, for decades, so it is what allows aircrafts to be produced over and over again. Sometimes the recipe can be slightly improved, but it is preserved in its ability to create an aircraft that can fly safely and swiftly. Thus, aside from all the various things that the factory contains, there is a particularly important piece of information, which is copied from generation to generation, in order for aircrafts to be produced to the factory's standard. Losing this recipe would make the factory fade away. And even if an alien civilisation were to find the remains of the aircraft factory, in order to make it functional again, they would have to find the recipe or figure it out themselves. The recipe is the abstract catalyst, and it is made of knowledge.

In order to be compatible with the physics we know, the recipe must have the form of a *sequence, or combination, of elementary steps*. To see why, we must understand that the laws of physics do not contain any protocol to preserve or create complex entities such as aircrafts, or even wings or things of that type. Leave an aircraft for a while in a desert, without being repaired or checked, and it will soon become unfit to fly. As I said in chapter 1, the only things that the laws of physics preserve directly are elementary components and interactions, and elementary symmetries. They are "no-design" laws. Ultimately, physical laws provide only very simple types of transformations that can be performed reliably without there being a recipe—those corresponding to naturally occurring interactions. These are transformations that happen spontaneously, such as the oxidization of the coating of an aluminium surface, or the molecules of air in an oven heating up the surface of a cake. Such transformations are elementary steps that do not need further maintenance to be realised in a stable manner; in fact, they do not even need to be specified in the recipe, because they are implicit in the laws of physics. I have uncovered a regularity in the way abstract catalysts appear: they must be realised as a recipe—a sequence of elementary steps, nonspecific to the output, each of which does not require further explanation and can be considered as a direct consequence of physical laws.

The fact that the recipe must have elementary steps, each of which is itself compatible with the laws of physics, also explains another feature of all catalysts. In biology, it is called 'the appearance [better: illusion] of design'. It is the trait of having several different parts all interconnected with one another, each one of which is dedicated to a particular function—that is, dedicated to realising a particular substep of the recipe. Something being dedicated for a particular function means that if you change it slightly, it no longer meets the criterion of being able to perform that function. Therefore, whenever you see something with the appearance of design that can last a long time, you can rightly assume that some abstract catalyst is contained in it.

I have said that most transformations that are possible in the physical world must occur via a catalyst that enables them. I have also noted that all catalysts must contain an abstract catalyst, which is itself made of knowledge, as introduced in chapter 1. Knowledge is defined entirely via counterfactuals: it is information that is capable of remaining instantiated in physical systems. Unlike most definitions of knowledge, the good thing about this one is that it does not depend on there being a knowing subject. This way of looking at knowledge, based on counterfactuals, breaks with a long-standing tradition, which regards it as a chiefly anthropomorphic, subjective concept. That tradition says that knowledge requires there to be a sentient being, such as a human, to exist. Knowledge, in other words, would exist only in minds. According to this idea, knowledge appears to be subjective. Something like that is very remote from the laws of physics. In contrast, knowledge as defined here, as it occurs in abstract catalysts, is sharply different. The main two differences are that this entity is objective (it can exist irrespective of whether there is a knowing subject) and it is a possible subject for physics. As I said, those who know the philosopher Karl Popper will recognise the chief features of his epistemology in the characteristics of this concept. However, with the Science of Can and Can't, I have related knowledge to physics, something beyond the domain of epistemology, which we can do because we are using counterfactuals. In physics, there have been several discoveries of new types of stuff. For example, at some point it was realised that all engines use some type of energy (heat) and transform it, partially, into some other type of energy (work). It was then natural to wonder how to express laws about these two types of energy—and that gave rise to thermodynamics. Likewise, in this case, it is natural to wonder: Can knowledge be created? Can it be destroyed? Can it be transformed? This problem is a deep one, and it is only partially solved so far. The Science of Can and Can't provides an objective handle on knowledge; it gives us tools that may one day be used to answer these questions fully.

As I write these pages, I am sitting in a café in the centre of Istanbul, several metres above ancient remains, admiring the rooftops and the minarets and the domes, drinking a tea whose serving ritual has been around for centuries, and observing a noble Persian cat curled up on a crimson sofa. With the Science of Can and Can't, I see that these things are all connected. Their resilience is the by-product of some abstract catalyst, which has the ability to be copied from generation to generation. With this new perspective, we have a different angle on theories that deny that knowledge could be anything substantial, or of scientific interest, on the grounds that it might be associated with an anthropocentric attitude. There is nothing anthropocentric in abstract catalysts. Their capacity to enable transformations is objective. In fact, knowledge and knowledge-creating entities are singled out as significant properties of our universe; but this is not, as in religious explanations, because of some dogma; it is because of a physical explanation. Knowledge is a particular property that matter can have in our universe—which exists when abstract catalysts are there; and it is fascinating to study its regularities: how it comes into existence, how it evolves, and whether it can be sustained and grow indefinitely. And this becomes, through this approach via counterfactuals, a problem for physics. Maybe one day we will be able to solve it.

The Wind Rises

The large villa housing the REBUS aircraft company in Trieste was immersed in silence. The ink-like darkness of a moonless night permeated everything, other than a small window in the ground-floor hangar. Inside the hangar, an oil lamp illuminated the surface of a mahogany table covered by books, tools, and sketches. A young woman was sitting cross-legged at the table, thinking. Giulia—that was her name—was looking intently at what, with any luck, would be her first repair job as the REBUS chief engineer. It was a small, agile aircraft that had suffered a serious engine failure. The name of the aircraft, KIKI, was painted in crimson capital letters on both sides.

The lamp was projecting long, trembling shadows on the walls; the aircraft resembled a metallic beast, frozen in its sleep. Giulia sighed quietly, staring at it. What made her uncomfortable was not the idea of having to repair a gem of that beauty. It was the thought of having to do it in such a hurry. The deadline was the following day; she had only a few hours left. In fact, there would have been plenty of time had she started the work early enough. Konrad A. Eckhart, the pilot and owner of the plane, had commissioned the repairing job exactly a month before. The aircraft had been delivered by a special truck; Konrad came a few days later to discuss the details of the repairing schedule. The engine failure had happened while he was flying. He had managed to achieve a brilliant emergency landing in the fields near Trieste. In fact, the aircraft had narrowly missed a vast flock of sheep, stopping its race in the middle of a green pasture. Everyone was still talking about the event: aircrafts were not that common in 1921, let alone in the rural areas where Konrad had accidentally landed.

Konrad was a character. He spent most of his time in the air, flying aircrafts of different kinds; when he was not flying, he would vanish from the inhabited world and find refuge on an island in the Adriatic to work on his deepest interests: theoretical physics and probability theory. He was also notorious for being curt and parsimonious with words. This caused people in Trieste to whisper and gossip almost constantly about Konrad, his private and public life, and his adventures.

Having heard all sorts of bizarre stories, Giulia was surprised to see how unimposing and gentle Konrad actually was in person. He showed up in his usual inconspicuous apparel—including that *dark green scarf he wore on every occasion. His eyes were a particular mixture of grey and green hues—they reminded Giulia of two calm mountain lakes. Being reserved and quiet herself, she felt immediately at ease with Konrad's short and focussed sentences. He asked for the repair to be done as quickly as possible, strictly by the end of the month. He spoke an almost faultless Italian. Giulia surprised herself, focusing on his nearly imperceptible accent as they were talking. Was it French? British? Polish, perhaps? Too faint to be able to tell. Whatever the accent, his words had been crystal clear. Konrad needed the aircraft by the end of the month to confront his rival Vincenzo Scarlatti in a flying competition in Trieste.*

For Giulia, this was one of the most exciting jobs she could have been offered at that stage of her new career. She had just completed her apprenticeship as an engineer and a pilot. She had been trained by her father, who had founded REBUS a while back, long before aircrafts started to become something worth investing in. At the time of Konrad's visit, Giulia had just inherited the company, after her father's sudden death. He had been trapped in a huge fire a few days before, while working in the garage of another company he was consulting for. As Giulia was scattering the ashes on the day of the funeral, she felt as if the wind was taking away all her ability to enjoy anything in the world. All those distant and close relatives moving about her, trying to be helpful, expressing their sympathies, produced in Giulia an extreme sense of loneliness—a sore, hopeless, and seemingly endless feeling of loss.

This is how Giulia managed not to repair Konrad's aircraft for a whole month. She was deprived of any sort of interest for aircrafts—or for anything else, in fact. Her main desire was to disappear somewhere in a deep forest and be alone, for a long, indefinite time. But she couldn't. Among other things, she felt she ought to do, there was that promise she had made to Konrad, to repair his aircraft.

So instead of disappearing, Giulia persuaded herself to show up in her workroom every day. She made it a point that she would stay in the workroom for as long as possible, and that she would think. Initially, what she did was not quite like thinking. It was more like daydreaming. She would let her mind drift, gently, unconstrained, in what felt to her like a grey haze.

She felt very much like a famous explorer whose fascinating memoir she had read a while ago. The memoir was an account of how the explorer managed to get sight of a rare species of red fox. He had to wait, and wait, and wait, in a most uncomfortable still position, for days, in the freezing cold of the Sierra Nevada mountains, until finally a vixen with three cubs made its appearance in the white, snowy woods. Like the explorer, Giulia patiently waited in the workroom for her real self to show up.

Eventually her real self made an appearance, sudden and furtive like a fox. It happened at a most unexpected point, while she was browsing through her father's detailed notes about engines and aircraft designs. Inexplicably, Giulia found herself completely focusing on their content, rather than on missing her father. Hard to say how that shift of focus happened, but every day after that she spent many more hours focussed only on browsing his richly illustrated notebooks, reading excerpts, noticing details of her father's drawings, thinking of ways of combining designs. In the end, she started thinking at full speed about creative ways of repairing Konrad's aircraft.

A few evenings away from the deadline, Giulia finally had the key idea. At first it was just a faint shadow in her mind, which she deliberately did not chase. She let it grow into something more definite, until it became a clear thought. Having an idea, after such a long time of emptiness, felt like having stumbled upon a momentous discovery. The idea involved an ingenious way to combine details of the engines of two different models with that of Konrad's aircraft, to make it more responsive and efficient. At that point, Giulia knew it was time to go back to working actively. This is how she ended up in the garage, so late at night, the day before the deadline. Despite being annoyed that she had to do everything in a rush, she was quite excited.

At once, she grabbed the tool kit her father had left her. Giulia's hands were familiar with the tools, and touching them after a long time—feeling their weight, the texture of their different types of surfaces—evoked distinct memories of her apprenticeship.

Knowledge

There was something in doing things with care, love, and skill that made Giulia profoundly happy. Fully responsible for the work she was doing, she brought it forward steadily, with focus and drive. As the hours passed by, Giulia was operating methodically, checking carefully all the bits of the engine. She was proceeding solely on the base of the information she had in her mind, gained over several years of apprenticeship, now perfected by her recent idea. She could almost feel the information in her brain guiding her hands while tightening screws and flattening bits of the metal surface. The pleasure of working was flowing steadily. She was acting on the world, on the account of her knowledge, improving it.

When the first light of dawn filtered through the window, it reflected off the polished metal surface of the repaired aircraft. The glittering fuselage looked like a jewel in the early-morning sun. Giulia suddenly remembered her father, and something he had once said. Something about the knowledge in their company being resilient and having the capability of perpetuating itself despite the fact that physical things have a finite duration and do not last. She had never quite understood what he meant, until that moment.

Later in the evening, at sunset, Konrad was saying good-bye. Despite his parsimony with words, he found a way to praise Giulia's work. Giulia thought she even saw the ghost of a smile make a swift appearance somewhere deep in his eyes.

After a few more security checks, Konrad took off effortlessly. Giulia stood for a long time on the balcony overlooking the take-off track, observing his aircraft ascending rapidly into the sky. In the distance, she could see restless wave patterns mixing cyan and grey hues on the sea surface. A mild evening breeze was coming from inland. It felt to her as gentle and warm as her father's hug.

6.

Work and Heat

Where I discuss the **conservation of energy** as a counterfactual principle about impossibility; three different kinds of **irreversibility** in physics—statistical, forgetful, and counterfactual; where I provide a **counterfactual second law**, based on an exact distinction between work and heat; and where you encounter the **universal constructor**, a machine that can perform all transformations that are physically possible.

Odette was tiny, graceful, and made of plastic. She was a ballerina—the centrepiece of a music box that I received as a present on my fifth birthday.

The logic of a music box is simple: you wind it up, it plays a tune. As I discovered later, the tune of my music box was an adaptation from the music that opens *Swan Lake*. It was a delightful collection of tingling notes.

As the music was playing, Odette danced on her wooden support, undergoing continual gyrations. She could twirl slowly and gracefully in her elaborate white dress for as long as the mechanical charge lasted.

The most intriguing piece of the music box was not the tiny dancer, though, nor the tune. It was the internal, clockwork-like mechanism. One could turn the box upside down and observe its intricate details through a conveniently transparent glass bottom. The brass key at the back of the box could wind up a coiled spring; the spring was connected to a set of gears that set a cylinder in motion. The cylinder had a pattern of pins on its surface that plucked the teeth of a steel comb, each of which produced a specific note; the pattern on the rotating cylinder coded for the tune. This inner mechanism was the most intriguing part, because it translated something material, the mechanical motion of gears, into something airy and abstract: a melody.

The reason why that translation happens resides (as I expect you have guessed) in a counterfactual property of physical systems. Let me explain how it all works. The music box contains an intricate mechanism, whose elements are necessary for the box to translate mechanical motion into music.* However, one more thing exists that is of the essence but invisible to the naked eye. It is the mechanical charge. The mechanical charge is the fuel that sets the whole mechanism into motion.

---

*To be precise—into sound (vibration of the molecules of air), which is then turned into music by our brain.

Where does that motion originate? At first sight, it comes from the user, who turns the winding key. However, if you go a little deeper, this seems to be the start of an infinite regress (which has appeared repeatedly in this book as signalling a problem in the traditional conception's explanations). There is no end in sight if you go down this line of enquiry, because you could ask the same question of the motion of the user's hand. Where does that originate? Also, innumerable other mechanisms could equally well wind up the box. For instance, a small mechanical engine attached to the winding key could do as well as the user's hand. To understand, you need to ask a more fruitful question. What is the mechanical charge made of? What kind of stuff powers the music box when you wind it up?

The stuff that charges the box is what physicists call *energy*. The term comes from a Greek word popularised by Aristotle whose original meaning in Greek is 'capacity to work'. Nowadays, what we call 'energy' in physics is even more sharply defined than in Aristotle's time. It is an abstract property of physical systems that must be subject to substantial constraints. As we know, the most important of these constraints is the law of conservation of energy, requiring that the energy of a system can be changed only at the cost of changing the energy of some other system by the same amount.

The law of conservation of energy is about counterfactuals. For it requires that it be *impossible* to change the energy of a system without any side effect. Given that *all* laws of motion must conform to the conservation of energy (those already known, and those yet to be known), the conservation of energy is more general than any specific dynamical law. Also, it is intended to apply to any system in the universe. It rules minuscule particles, such as electrons and protons; heat engines that propel aircrafts and spaceships; and the mitochondria powering our cells. It applies to anything and everything that has an energy, independent of its scale and size.*

This seemingly innocuous requirement has sweeping consequences. First, it implies that the energy of a system cannot increase (or decrease) unless energy is supplied from (or absorbed by) something else. This allows us to account minutely for the whereabouts of energy as it transits from one system to another. When it escapes from here, it must have gone over there—it can't disappear or pop into existence. For example, the music box cannot get charged spontaneously without something else providing the energy (the winding of the key). Likewise, the battery of a smartphone does not get replenished if the device is switched off and not connected to the power supply.

Still, the energy of a system can change by interacting with some other system (which will absorb or give up energy in turn). The conservation of energy still requires that the energy of the two systems together not change, if they are isolated. The exact

---

*In general relativity, the law of energy conservation must be upgraded to apply to a more general quantity, whose details are not important here. But *mutatis mutandis*, the considerations I make in this chapter apply there, too.

amount of energy added to one system must be subtracted from the other so that their overall energy does not change.

The mechanism in the music box, which converts the mechanical charge into a tinkling tune, is allowed by the conservation of energy and conforms to it precisely. When we rotate its winding key, we add energy to the box. The rotation of the key compresses the coiled spring, which thereby gains energy. At the end of the process, if there are no losses along the way, all the energy spent to turn the winding key ends up inside the spring. The spring is a repository of energy—just like the battery in a smartphone. It connects to the other mechanisms enabling the cylinder to rotate and then to pluck the teeth of the comb. When we open the box, the spring is activated and gradually expands. It loses energy by providing it to the rotating cylinder. The rotating cylinder then conveys the same amount of energy to the metallic comb, which vibrates, reproducing the right notes at the right time. You can extend this reasoning to any other system involved in the story of how the music box plays its tune. For instance, each tooth on the steel comb, when vibrating, sets into motion molecules of air that surround it. The notes one eventually hears are caused by waves of vibrating molecules of air that impact the eardrum with a specific frequency of vibration. In turn, that mechanical stimulus turns into an electric signal when processed by your brain. The air waves carry energy, too. So do the brain signals (but the latter have been amplified using some of the energy from your body).

The conservation of energy implies the startling prediction

that, at each step along this chain, if one accounts for all the systems involved, one shall find overall the same amount of energy initially given to the winding key. Such is the inescapable accountancy set by the law of conservation of energy; and it is based on the counterfactual property that it is *impossible* to change the energy of a given system (without side effects being produced).

The music box embodies another remarkable counterfactual property of energy. The energy stored in the box's spring is *interchangeable* with the energy in the clockwork-like mechanism, or with the rotational energy in the cylinder, and so forth. You can store the very same energy in all sorts of diverse systems, independent of most of their particular physical details, and you can transfer it freely from one to another. Energy can flow from the spring inside the box to the rotating cylinder, to the comb, to the vibrating molecules of air, et cetera. The systems in the music box are interoperable: energy can be transferred from one to the other, regardless of most of their particular details. You have already encountered a similar counterfactual property, the interoperability of information, in chapter 3. There, the information media were all interchangeable with one another because they all had the same counterfactual properties. Does a similar interchangeability hold for all systems that can embody energy?

Interestingly, the answer is no. Only some of the systems that can embody energy are interoperable, in the sense that they are all interchangeable as far as energy exchanges are concerned. These systems share the same counterfactual properties, which I shall tell you about later in this chapter. Other systems that can contain energy are not interchangeable. That is due to the second law of thermodynamics.

The second law distinguishes between two types of energy transfers—and the distinction is rooted in counterfactuals. One type of transfer is reversible: one can use these energy transfers to perform work on a variety of physical systems, such as the brass winding key of the music box; or a flywheel; or a piston; and then undo the transfer completely, retrieving the energy in full, with no irreversible losses. The systems supporting these energy transfers are fully interoperable, just like the systems in the music box.

The other type of energy transfer is irreversible: once the transfer happens, it cannot be fully undone; part of the capacity to do work is lost along the way. A classic example: the brakes on a bike. When you brake, you apply resistance against the rotation of the wheel. The wheel and bike initially have a certain energy, and they come to a stop: the brakes and the wheel itself have heated up. Why? The energy that was in the motion of the wheel and bike has now gone into the thermal motion of the molecules composing the wheel and the brakes, and it is practically irretrievable. (I shall revisit 'practically' later!) The same holds for the energy of the vibrations bringing the tune of the music box all the way to our ears. Once the music is heard, it is very hard to bring that energy back into the box's mechanism.

The chemist Peter Atkins has frequently said, in his masterly books about the foundations of thermodynamics, that work and heat are not 'substances' (just as information is not a substance). They refer to *modes* of transfers of energy. I call the reversible transfers *work-like* (the transferred energy that can be reused ad infinitum, to initiate or to stop controlled, ordered kinds of motion); I call the irreversible transfers *heat-like*. The second law requires some energy transfers to be heat-like: once they happen, it is impossible to recycle some of the energy involved in them. That energy can no longer be used fully for a work-like transfer; only some of it can.

The second law of thermodynamics operates in a rather impressive way. In tandem with the principle of the conservation of energy, it provided the theoretical foundation for heat engines, which powered the incredible progress that occurred during the industrial revolution. But when it comes to explaining exactly what the second law says about the physical world, the issue is not as clear as for the conservation of energy. It is so complicated and subtle that physicists over the decades have proposed numerous inequivalent formulations of the second law. Each has its notion of heat-like and work-like transfers of energy, and they are different from each other! Still, all these different formulations concur on a few striking consequences concerning the thermodynamics of heat engines. The second law is thus a pillar of the edifice of theoretical physics, but we are not quite sure what it means.

Whatever it means, physicists are quite happy with it because, in the domains where it applies straightforwardly, it has been incredibly successful. Recently, however, technology has brought us closer and closer to probing domains where the second law becomes problematic or ambiguous—for instance, in the development of engines operating at the nanoscale, the scale that characterises, say, the interior of a human cell or the circuits in classical and quantum computers. At this scale, the distinction between what is heat-like and what is work-like becomes blurred. If we are to use the second law to describe these systems with the same accuracy and efficiency as we do for macroscopic heat engines, then we need to sharpen the law.

Luckily, as I am about to reveal, counterfactuals provide the key tool to crack this problem. Once one recognises that the second law is about counterfactuals, the path to making it exact at all scales becomes clear. The logic is elegant and follows closely the approach I used to solve the problem of how to express information exactly within physics in chapter 3.

First, you need to understand what the problem with the second law is. In short, the problem is that the second law requires some irreversibility. Incidentally, that is also why it is so fascinating. Irreversibility is at the core of various phenomena that are ubiquitous in physical reality: the birth, development, and death of organisms; the growth of complexity in the biosphere; the increase in sophistication within our civilisation; the creation and destruction of knowledge. The irreversibility requirement of the second law brutally clashes with the laws of motion ruling the elementary constituents of matter. Remember? I said in chapter 4 that the laws of quantum theory are reversible: if they allow for a transformation, the reverse transformation must also be possible. The laws of general relativity (the other most accurate description of physical reality we possess) are reversible, too. If there is a trajectory that takes a system from A to B, there must also be one that takes it from B to A. Microscopic constituents of matter must operate in this reversible manner because they obey these laws of motion. The problem, then, is: How can the second law require that some energy transfers are irreversible and be compatible with the reversibility of the laws of motion?

The contemplation of the second law (irreversible) and the laws of motion (reversible) feels like looking at one of those optical illusions called autostereograms. If you look at the image in one way, all you see is a flat, 2-D pattern (reversibility). However, if you look at it from a different angle, suddenly you see a 3-D figure emerge (irreversibility)! How can 2-D and 3-D images coexist in a coherent picture? For the optical illusion, we know it is not possible to hold both images in your mind at the same time; but you can hold an explanation for how they are both there, on the page. Likewise, is there a unique picture of physical reality that can reconcile reversibility and irreversibility? Physics does not yet have a definite answer to this question, let alone a unique one. There are a few proposed answers, but each is still controversial. Ultimately, it is because theoretical physics is trapped in a world without counterfactuals. With counterfactuals, one can reconcile the reversible description of the laws of motion and that of the second law, at all scales.

To see how, I shall first take a closer look at the irreversibility of heat-like transfers. Imagine a playground; and in that playground, a seesaw. A seesaw consists of four main systems: two seats (of approximately equal mass), and a rigid, long, sturdy bar joining them, slotted on a pivot placed at its midpoint. We need two children to play: a child sits at each end so that one goes up as the other goes down, by gravity, depending on which child weighs more. Added fun comes if the children attempt to go as high in the air as possible as they take turns pushing their feet against the ground.

Let's simplify the picture a little. First, imagine that, instead of the children, there are just two springs firmly secured to the ground underneath each end of the bar. Also, imagine that both ends (call them A and B) have the same mass. The seesaw in the neutral position corresponds to the bar being perfectly horizontal above the pivot: still. Now suppose you push one of the ends, say A, in the upwards direction. Up it goes, while the other end (B) goes down and compresses the spring as much as allowed by the conservation of energy. The energy given by your push is now transferred entirely to the spring. Then the compressed spring gets decompressed by its completely elastic nature, thus giving back the energy to the B end of the seesaw, which therefore goes up. End A goes down, compresses the spring in turn, and so on.

The motion continues forever in this fashion if no other dissipative interactions intervene. The transfer of energy in this system is reversible: energy goes, reversibly, from the spring, to the A end of the bar, to the B end, and finally to the other spring. In that oscillatory motion, energy is reversibly transferred from one system to the other. All exchanges are work-like.

I'm guessing that by now you may be a little puzzled. You were trying to follow my argument, yet each of its steps contradicts what you expect to experience in reality. If you try to use a seesaw as I indicated, at best the bar will bounce back on the spring once, and after a few damped oscillations it will come to a stop. Nothing like a continual oscillation!

Your perplexity is entirely justified. My explanation does not contradict your intuition about reality. The critical difference between the real-life situation and our imaginary one is that I imagined a case in which no other interactions exist in addition to those specific to the seesaw mechanism. It is an idealised case: it is, in principle, possible to get arbitrarily close to this scenario, short of perfection, but it is far from what happens in reality. Indeed, around the seesaw, there are several sources of disturbance, unspecific to the seesaw motion, which dissipate the energy at each step.

For a start, there are countless molecules of air continually bouncing off the bar as it goes up and down. Then, the spring is also not perfectly elastic: the energy transmitted to it by the bar is not entirely returned when the spring gets decompressed. Some energy goes to waste, away from the seesaw oscillations—for example, it is absorbed by the atoms of the spring, which increase their internal vibrational energy. So there are several more interactions to take into account to explain where the energy goes in a real-life seesaw. All these interactions take a little energy away from the combined motion of the seesaw and the springs. That is why, if a real-life seesaw is set into motion and then left alone, it eventually comes to rest. Overall, though, energy is still conserved. When the seesaw comes to a stop, the energy given by the push is stored in the air molecules, and in other particles inside the spring and the bar. They are all a little warmer (more energetic) than before.

Here is the point where irreversibility creeps in, in a world where all elementary interactions are reversible. Once the energy has gone into molecules of air and vibrational motions of atoms, it becomes tough to bring it back, in practice. Such kinds of energy transfers are those the second law labels as heat-like, generally regarded as irreversible—just like in the case of the brakes or the music box.

Watch out now! Here is the point where the water becomes muddy. It is one thing to say that reversing heat like energy transfers is impossible 'in practice'. That means that it is hard, but it is still, in principle, possible. It is entirely another thing to say that reversing them is impossible—that it is categorically forbidden, just like creating energy out of no energy is. The fundamental question for the second law, then, is this: Is it just that reversing certain kinds of energy transfers is highly impractical, but in principle possible, or is it that it is strictly impossible?

Here the very status of the second law as a law of physics is at stake. In physics, as I mentioned in chapters 1 and 2, the most fundamental laws are exact and universal—like quantum theory's or general relativity's, or like the conservation of energy. These are exact because they say how a physical system *must* behave, not how it approximately behaves plus or minus some tolerated deviations. Also, they are universal because their domain of applicability is the whole universe. If the laws are not universal, then they have to come with a specification of where they stop working. As an analogy, think of a manual for some kind of appliance—say, your smartphone. First, you want the manual to tell you exactly what to do in each possible situation, not what to do 'more or less'. You also want it to indicate precisely what models the manual holds for, so you know when to depend on it and when not. The laws of physics are tentative manuals for the use of the universe. The second law, as we know it, works well in some domains, but it is not exact and it does not have a well-defined domain of applicability. It is a great manual for some macroscopic systems (e.g., locomotives in trains), but we do not know if it works and what it says for systems that are much smaller than that.

If reversing heat-like energy transfers were just very hard to achieve, but ultimately possible, the second law would not really be a fundamental law—in fact, there would be no reason for physics to distinguish between work-like and heat-like; there would be no irreducible irreversibility. All energy transfers would be work-like (i.e., reversible), only a little harder to achieve in the reverse direction compared with the forward direction. The second law, and its prescribed irreversibility, would just be the description of our current technological limitations—which of course are not fundamental. These limitations can be improved upon by investing enough resources into it.

On this question of whether the second law is fundamental, physicists are currently divided into two main camps. One camp says that it is not fundamental: there are only reversible laws, governing the microscopic interactions of particles. With enough technological resources, the reversible dynamics could always bring all energy back to where it came from, and one could then reuse the energy to do work. This would imply that the limitations imposed by the second law on heat engines are just a rule of thumb, telling us that it is hard in practice to reverse specific interactions. But these limitations could be lifted, ultimately, by improving our technology. According to this camp, the second law does not need to appear in the manual for the universe.

The other camp claims that the second law is fundamental: that there can be a formulation of the law that is universally true, and still compatible with the reversible dynamics, at all scales. Can this be so? Various paths to reconciling irreversibility with reversible laws of motion have been proposed to support this idea. None of them really works to the end of creating a universal, exact law; they ultimately all concede that irreversibility only appears as some sort of approximation; but that it is not fundamental. Only one of these paths is based on counterfactuals, and as I shall explain, it is the only one that has some potential to be successful in this endeavour. I shall go into a little detail about these paths because they are smart ideas, even if they end up with mixed success in regard to producing an exact second law. Understanding the other approaches is indispensable to grasping the superiority of the approach with counterfactuals.

One way to define heat-like, irreversible energy transfers is statistical—it uses the branch of physics that goes under the name of 'statistical mechanics'. This path leads to a formulation of the second law, stating that irreversibility in certain physical systems must occur with high probability. It leads to a *statistical* second law. Here is how it goes: Suppose you leave a glass of iced tea out in your garden on a midsummer day (when it is not raining—in countries like England, where I live, this hardly ever happens, but we can use our imaginations). What we all expect is that the iced tea will gradually warm up, acquiring the temperature of the environment. The second law, formulated in the statistical form, says that the configuration where the tea has equilibrated its temperature with the environment and stays there, unchanged, is the most probable of all possible configurations. Other configurations (e.g., where the cup cools down even more) are still permitted by the dynamics but are highly improbable. In this formulation, if you were to repeat this experiment many, many times, in the overwhelming majority of cases you would see the equilibration of temperatures occurring. But not always!

The brilliant physicist Ludwig Boltzmann sharpened this law by formulating a criterion for what processes are more likely to occur in a physical system (i.e., the glass of iced tea) interacting with an environment much larger than it (i.e., the garden). According to Boltzmann, the most likely physical processes tend to increase or leave unchanged a property of the system, its entropy. The entropy is a function of the state of motion of the particles in the system. It is often loosely linked to the degree of disorder of the molecules—the more disordered they are, the higher the entropy. (It is an imprecise connection, in general, but for our purposes, it will do.) Following Boltzmann, the second law, in this statistical formulation, says that the most probable states of a system in contact with a large environment are those that maximise the entropy of the system or leave it unchanged. Under this formulation of the second law, heat-like energy exchanges of a system are those leading to an increase of entropy via energy exchanges with the environment, just like the heating up of the glass of iced tea in the garden, or the dissipative bouncing of the seesaw in our other example.

The statistical-mechanical law cannot even aspire to be exact or universal. The configuration maximising the entropy is not the only one guaranteed to occur; it is the 'most probable'. All other configurations can still occur, but it is not said when and how, only that they are 'less likely'. The fundamental reason why the statistical second law cannot be exact is that the dynamics regulating the exchange of energy between the iced tea and the surroundings are reversible. From the microscopic point of view, here is what happens when the tea is placed in the garden and left there. The molecules of tea are much less energetic than the molecules of air because the tea is colder than the air. At the surface of the liquid in the glass, the molecules of air interact with the molecules in the glass. You can think of these interactions as collisions. By collisions, the particles of tea acquire a bit of energy from the particles in the surroundings. One of the possible behaviours does lead to the state where the tea is no hotter than its environment, and the environment is no colder than the tea: their temperatures are the same, and they do not change any more thereafter. In order to have irreversibility, the tea should be able to reach this state and then get stuck on it indefinitely. But it is not what the laws of motion imply. In fact, the laws of motion imply something different—a rather impressive fact: if you waited long enough, the tea would eventually go back to its original state, which means to its initial temperature. This might require you to wait for a long, long time—longer than the resilience time for the glass. However, there will be a time when the laws bring every particle, including those in the glass, back to some point visited a long time before. The recurrence is an astonishing prediction of a theorem that we owe to a fine and subtle mathematician (and engineer!)—Henri Poincaré. His theorem says that all reversible laws have the property of bringing the system back to its original state (within an arbitrarily small distance from it, to be precise), in some finite time. Of course, it could be a very long time, but it is finite.

Given that the microscopic laws of motion are reversible and conform to Poincaré's theorem, they leave no space for irreversibility of this kind. Therefore, the statistical form of the second law can predict only irreversibility with high probability—not with certainty. It will never be an exact law, like, say, Newton's law is. This path to irreversibility does not lead us to a fundamental, second law—only to one that is, at best, approximate. Given our aim, we need tread this path no further, and we shall now move to another.

Let's follow another path towards reconciling the irreversibility implied by the second law with the reversibility of our dynamical laws, such as general relativity and quantum theory. The second path, also deeply problematic, derives irreversibility by forgetting some details of the underlying reversible dynamics. If this seems to you like a cheat, it is! Imagine a reversible motion, set up as follows: a coloured lamp changes from red to green, from green to blue, and from blue to red. The lamp's behaviour is reversible: if you look at the lamp at a certain point and try to infer what state it had immediately before, you will be able to guess with certainty what the state was. If it's red now, it was blue before; if green, it was red; if blue, it was green. You will then be able to reset the lamp to its previous state, which is why one can say that the behaviour is reversible.

Now imagine you are wearing glasses with lenses that merge blue and green into the same colour—yellow. You would then be able to distinguish red from blue and red from green, but not blue from green. In this case, the distinction between green and blue has disappeared as far as you are concerned. So the dynamics of the lamp will not be reversible, from your point of view. Seeing the lamp showing yellow, you will not be able to infer with certainty what the state of the lamp was before. It could have been green; but also red. The underlying dynamics of the lamp are still reversible; but unless you remove the glasses, you will not be able to reset the lamp to its exact previous state. This example can be generalised: the impossibility to distinguish certain microscopic states can result in irreversibility even if the microscopic laws are reversible.

This kind of 'forgetful irreversibility' is the basis of a different form of the second law. It states that the irreversible behaviour occurs only when one looks at specific macroscopic properties of physical systems—meaning only when you forget about some details of the dynamical states of a system. But this also leads to a non-exact, non-universal law. There is no criterion to choose what to forget: why should one forget about certain aspects and not others? In current accounts, no one seems to have an explanation for what to discount. In our example, this corresponds to the fact that what kind of coloured lenses one puts on is arbitrary. Arbitrariness leads immediately to subjectivity. Some observers could forget some details about a given motion; others may forget about others. For some, a process could look irreversible; but for others, it may look reversible. And there would be no law of consistency between what all the observers see, given that the choice of what to forget is arbitrary for each observer. Treading this path, we conclude that irreversibility is possible only thanks to arbitrary limitations that confine your attention to only a selected subset of properties of physical systems. But why should one do that? This parochial limitation cannot be the foundation of fundamental irreversibility, because it is arbitrary and subjective; it can always be corrected by remembering what was temporarily forgotten, going back to a reversible state of affairs. Sweeping things under the carpet is never a good way to approach problems. We shall abandon this path to irreversibility because it also does not lead to a fundamental second law.

Just like Goldilocks in the three bears' house, after trying two paths that do not work, we land on the third path to irreversibility— based on counterfactuals. This path is not 'just right'—it, too, has problems—but it is more promising than the other two (which are instead based on the traditional conception of physics).

To set off down this path, let's trace the same conceptual steps that led the superb physicist James Joule to perform a crucial experiment in the early days of thermodynamics. What he conjectured and verified experimentally was that while it is *possible* to heat up a volume of water by stirring it only mechanically, it is *impossible* to cool it down by those same means. You can see that I am now talking about counterfactuals: the language regarding possible/impossible transformations belongs to the Science of Can and Can't.

Let's return to the glass of iced tea. (Joule would have preferred a glass of beer, as he was also a brewer, but I shall stick with the tea; it can't do any harm.) Imagine you stir the tea vigorously, mechanically—say, with a spoon. This stirring provides the molecules of water with more energy. Imagine that the glass is somehow perfectly isolated from the rest of the environment, and no energy other than the energy of the stirring can be exchanged with the environment. What you will find out is that the tea in the glass ends up in a hotter state at the end of the stirring. On the other hand, no matter how hard one tries, the temperature cannot decrease through stirring only. In this scenario, that transformation is impossible. (Of course, if the glass is not isolated, you can stir the tea to cool it down, by facilitating exchanges with the air in the environment. However, here I imagine the cup to be entirely isolated.)

This kind of irreversibility prescribes that sometimes a transformation (such as heating some amount of water) is *possible by mechanical means* ONLY (e.g., using the stirrer); but the reverse task is *not possible* just by using those same means (though it may be possible by other means). Work-like energy transfers are those corresponding to transformations that can be performed by mechanical means only, in *both directions.* Heat-like transfers correspond to transformations that are *possible* by mechanical means in one direction only but are *impossible* in reverse, using the same means (and nothing else). As you can see, this path to irreversibility is about possibility of certain transformations and impossibility of their reverse—it is about counterfactuals. This approach to the second law is due to Lord Kelvin and Max Planck. It does not talk about the most probable free evolution of a physical system in contact with an environment, or about what trajectories a system accesses once you discount some of its details.

The fascinating revelation is that this kind of counterfactual irreversibility is compatible with time-reversal-symmetric laws, without requiring any approximation! Because even under perfectly reversible microscopic laws, it is *possible* to have some device that can perform a transformation in one direction by certain means (e.g., there can be a machine, such as an automated stirrer, that heats up a liquid by mechanical means *only*), whereas it is *impossible* to have a device performing the task in reverse with the same means (e.g., cooling a liquid by mechanical means only). Crucially, reversing the laws of motion of the elementary constituents will not turn the 'forward' device (performing a transformation from A to B) into a 'reverse device' (performing the inverse transformation, from B to A). Even if the elementary constituents of both devices obey reversible laws, the forward device does not necessarily imply the existence of the reverse device! Even if their elementary constituents behave reversibly, you can have that the forward transformation is *possible*, whereas the inverse transformation is *impossible*. This irreversibility is different from the statistical irreversibility, which requires that the forward *trajectory* of a freely evolving system is overwhelmingly more likely than the reversed trajectory. It is also different from the forgetful irreversibility, where one trajectory happens, whereas others do not, only if one neglects some details of what is going on. The latter statements can be only approximate (based on probabilities or arbitrary neglecting procedures); by contrast, the statement that a transformation is possible and its reverse is impossible is *exact*—it involves no arbitrary forgetting or probabilistic approximations. This counterfactual path to irreversibility and to formulating the second law is exact.

However, as it stands, this approach suffers from a serious problem. It does not explain what 'mechanical means' are—its domain of applicability is therefore undefined. A stirrer qualifies as mechanical means. So do an ideal spring and a suspended weight. But does, for instance, an atom in a well-defined state of energy count as mechanical means as well? What about a current looping in a superconductor? Or a photon with a well-defined frequency? Without a criterion to decide what counts as mechanical means, the statement "This transformation is possible with mechanical means only, but its reverse is not possible with those means only" is exact but does not say anything specific about the universe. A second law expressed along these lines remains unclear; so it cannot be a useful addition to the manual for the universe. For it to be clear and useful, it needs to explain what 'mechanical means' are. Here is where the interoperability property, which I hinted at while describing the music box, comes in handy. The solution comes in beautifully, once more, through the counterfactual approach!

We can tackle the problem using the same logic adopted for information and knowledge. We have to express the characteristic properties of mechanical means. A good place to start is to notice that they are capable of undergoing work-like transfers among one another. What should they be like in order to have that property?

Having energy, of course, is not sufficient. Some amount of iced tea at a given temperature has energy in it, but it cannot be used as a mechanical means by itself. The thermal motion of its molecules would not, by itself, enable one to put into motion other systems, such as a wheel or a pendulum. But a compressed spring or a suspended weight can do so, just by being connected to the system that we intend to set into motion. The additional property characterising 'mechanical means' is a counterfactual one—they must display some kind of interoperability, akin to that of information media, described in chapter 3.

To express it, a good place to start is to consider the counterfactuals in common to well-known cases of mechanical means: I shall start with an example, and then proceed with abstracting a general rule. Incidentally, expressing interoperability for mechanical means will unexpectedly reveal a significant connection between systems that can enable reversible, work-like transfers and systems that can store information.

The property is elegantly expressed by resorting again to a variant of our seesaw example—involving two weights hanging on either side of a pulley at some height above the ground, as in the figure on the next page. The first important point to notice is that different heights for the weights indicate different values of energy for each, because I am imagining that (as in the seesaw case) a gravitational field is present, which means (following a simple logic based on Newtonian mechanics) that the higher a weight is suspended above the ground, the higher its (potential) energy. (If this sounds counterintuitive to you, think of the familiar case of water falling from a height in a waterfall. The higher the fall, the more energy the water carries with it.)

The weight on the left is weight A and that on the right is weight B. The two weights can be in a variety of different positions, but there are mainly four configurations that matter here, as far as energy exchanges are concerned—represented in the figure. In one state, the weights on both sides, A and B, are at the same height; call this state (A(0), B(0)). You can think of A(0) and B(0) as respectively representing the level at which the weights are in this state. In this state, both weights have the same energy. Then there is the state in which the weight on side A is higher than A(0), while the weight on side B is at B(0); call this state (A(+), B(0)). In this case, weight A has more energy than B. Then we have the state where the weight on side A is even farther up and the one on side B is farther down, which we can call state (A(++), B(− −)). Finally, there is the state where the weight on side B is at a higher position

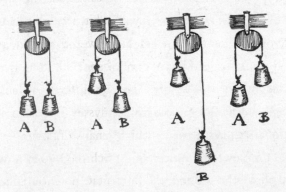

than B(0) (at some height h relative to B(0)), while the weight on side A is still at A(0); call this state (B(+), A(0)).

In this system of suspended weights, the following transformation is possible, with *no other side effects*:

$$(A(+), B(0)) \rightarrow (A(0), B(+))$$
$$(A(0), B(0)) \rightarrow (A(+), B(-))$$

I shall call this a *seesawing transformation*, to remind us of the alternating motion of the two weights. That transformation being possible means that one can set up a mechanism which, when given the weights in the state (A(+), B(0)), sends them to the state (A(0), B(+)); and likewise, when given the weights in the state (A(0), B(0)), sends them to the state (A(+), B(-)). One such mechanism is the pulley system I described above, with a suitable set of pins that can keep the weights from falling farther once they have reached the desired height (this can be done in a careful way so as to be frictionless). It is important to stress that *no other side effects* (involving other systems—such as other bits of the environment) are required, because it is what makes the weights in this example, and mechanical means more generally, special. The mechanism itself remains unchanged—which is what makes it behave like a catalyst (as I defined it in chapter 5: a system that can enable a transformation and retain the ability to enable it again).

So now it is easy to express what mechanical means are: they are all systems with different energy states having the crucial property that the seesawing transformation expressed above is possible. In the case of a general system, the states B(+), B(0), and B(−) refer to different values of energy for whatever system one considers, with B(+) being higher than B(0), which in turn is higher than B(−) (likewise for A). For instance, for an atom, they could be three different energy configurations where the outmost electron cloud has higher energy in the state B(+) than in B(0); and in B(0) it has higher energy than in B(−). Transferring energy from one atom A to another atom B, which corresponds to the transformation (A(+), B(0)) → (A(0), B(+)), would then happen via the emission of a photon from the atom A in the state A(+) and the absorption of the photon by the atom B in B(0). Similarly, for the transformation (A(0), B(0)) → (A(+), B(−)), where it is atom A that absorbs the photon emitted by atom B. The photon would act as pulley and ropes, joining the two weights in our example and making the above transformation possible with no other side effects on the two atoms.

I have characterised mechanical means with a counterfactual property: they are the physical systems with the property that the seesawing transformation defined above is *possible*. To highlight the fact that this characterisation embraces far more general things than mechanical means, such as springs and weights, I shall call systems that have that property 'work media'. They include weights and springs but also microscopic particles like atoms and qubits in particular states; but not glasses of water or cups of tea at a given temperature. The fact that they permit a seesawing transformation is the counterfactual property that singles out *all* systems that can undergo work-like, reversible energy transfers among one another. It is what we needed to complete the formulation of the counterfactual second law, to make its domain of applicability well defined.

A work-like transfer of energy is one that transforms a physical system from a state S to a state S′, and back again, requiring a change in energy on work media only. For instance, the ideal, frictionless seesaw implements a work-like energy transfer between the two weights on each side, because each of them qualifies as a work medium. On the other hand, if one of the two transformations (S→S′ or S′→S) is *impossible* to perform with side effects on work media only, the energy transfer is heat-like—that is, irreversible. So going back to the Joule example, heating up a cup of tea by stirring involves a heat-like transfer of energy from the surroundings, because it is not possible to perform the transformation in reverse (cooling) by mechanical means only.

The counterfactual second law, then, can be expressed (concisely and with no approximations!) as requiring that there must be heat-like transfers in the universe. In this form, the second law can be applied to all scales, independent of the kind of system; it is exact. Bingo!

The traditional, macroscopic second law was successful with macroscopic heat engines, such as those in trains and cars; but this extended counterfactual second law has the potential to apply to their nanoscopic equivalent entities. For instance, it applies to the nanoscopic electronic devices in your phone; to the qubits in quantum computers; to the natural and artificial molecular assemblers that operate at the scale of our cells. The definition of work media (that they are all systems on which a seesawing transformation is possible) is wonderfully general: it applies to a weight suspended in a gravitational field as well as to an atom with different energy levels available for its electrons, and it does not depend on the scale. What remains to be done in this case is to derive predictions from this extended counterfactual second law in the domains that the traditional formulations cannot cover. This kind of research, which requires a joint effort of both theoretician and experimentalist, is under way. If it goes well, it will provide us with groundbreaking technological outputs, which will harness the properties of microscopic systems in order to realise nanoscopic heat engines and assemblers.

There is a last tile to add to the mosaic I am constructing. Using the counterfactuals I described in chapter 3, you will find another far-reaching unification: work media are information media. It is because the seesawing transformation can also be used to distinguish the states A(0) and A(+)—you can convince yourself of this by noting the similarity between the seesawing transformation and the copying transformation defined in chapter 3.

In other words, systems that can be used to perform work-like transfers of energy must also be able to store information. They must have at least two distinguishable states (A(+) and A(0) in our notation) that can work as a bit. Energy states such as A(+) and A(0) that are usable for work-like transfers are distinguishable; they can store information. That's what distinguishes them from energy states enabling heat-like transfers—which need not be. The fact that any system usable to perform work-like energy transfers can also be used to store information is a profoundly unifying link between information theory and thermodynamics— the link between heat engines and computers.

This unification is not just elegant. It is also extremely useful in practice, just as Turing's theory of the universal computer was essential to develop the information technology that now sustains our civilisation. The path just trodden, connecting information and thermodynamics through work media, leads us to consider a fantastic possibility. There could a more general branch of physics, encompassing both information theory and thermodynamics, providing fundamental, universal principles constraining laws of motion that we know and that we do not yet know. Just as the theory of information led to the theory of universal computation, this theory I am envisaging could be the seed for designing a machine that generalises the universal computer, which scientists call the *universal constructor*. This machine was first conceptualised by the polymath John von Neumann. It has in its repertoire all physical transformations that are physically permitted—not just computations, but general constructions, including thermodynamically allowed ones (e.g., cooling down various systems), biological ones (e.g., self-reproduction and related biological functions), and much more. All in one single machine!

It can be thought of as the ultimate generalisation of a 3-D printer: when inserting an appropriate programme into it, and giving it enough rough materials, the universal constructor would construct out of them any system that is permitted by the laws of physics. The realisation of a universal constructor, though presumably very far in the future, could have epoch-making consequences—comparable in reach with those of the universal computer, which paved the way to the current information technology era. Formulating an exact theory of thermodynamics, as I have done in this chapter, is the first step necessary to construct the theory of the universal constructor, opening up avenues that will provide a radically new perspective on the physical world.

Labyrinth

There is a less known, yet gripping, story about Ts'ui Pên—the mysterious character invented by Jorge Luis Borges in the tale "The Garden of Forking Paths". I became aware of this story through conversations with the Earl of Kendal, an eccentric scholar who lived for a while in Oxford at the same time as me. He studied the history of printing in Europe from the Middle Ages to contemporary times. Being a bibliophile, he had several interesting stories to tell, collected from this or that book, and powerfully augmented by his imagination.

The Earl and I used to meet regularly in the richly decorated common room of University College, where he worked during his time at Oxford. He would come in at two o'clock, every day other *than Sunday, as punctual as the precision of his golden pocket watch. During our encounters, he would invariably conjure up some story about the most formidable characters or set of circumstances.*

One day he was showing me a particularly beautiful edition of The Aleph and Other Stories *by Borges that the Bodleian Library had just acquired. After contemplating the book for a while, we ended up conversing about Borges and his characters. At one point, he asked: "Did you know that Ts'ui Pên appeared earlier in a completely unknown story, documented to have appeared in another manuscript by Borges? Never published—it was lost after the writer's death."*

I did not know about that. In fact, the Earl appeared so keen to relate it to me that I decided to linger in the common room and indulge him. I have no idea whether the story that followed was a faithful account of real facts, or just the Earl of Kendal's invention, or a mixture of both. It is, however, a significant story about impossibilities. I shall recount it below, in all the detail I can recall, so that it can be remembered.

Ts'ui Pên, as Borges tells us, is a Chinese scholar who decided to leave behind all the pleasures of life to seclude himself in his house for thirteen years, to write a book (about an unspecified topic) and to build a labyrinth. Before that—and this is what the Earl's story is about—he participated in an exciting challenge, which he won. It, too, involved a labyrinth.

Ts'ui Pên used to visit several libraries in the vicinity of his mansion. They were open only to a few illustrious scholars. One day, Ts'ui Pên was casually browsing a book about mazes when he found a note between the pages, handwritten in crimson ink: "Here's a curious problem for a curious mind. Construct a unicursal labyrinth, with no bifurcating or joining paths, with only one entrance and only one centre, which it is possible to enter, but impossible to exit. I deem this to be possible. I challenge the reader to explain how."

There was no signature. On the page where the note was found, there was a beautiful illustration of a unicursal labyrinth: one that is made of a single path, with no bifurcations.

As the meaning of the note was sinking in, Ts'ui Pên felt a rush of excitement. He kept staring at the note, suddenly free of the tiredness he had experienced earlier due to a lack of sleep in the past few days of intense study. He set his mind to the challenge without hesitation. It sounded mysterious and profoundly fascinating.

He went home right away, locked himself in his study, and spent the whole day thinking about the note, sketching on his notebook possible ways in which a visitor to the maze could be trapped. The labyrinth could not have more than one path, the note said. It had to be unicursal. Once one reaches its centre, it should be possible to come out by merely reversing one's steps and following the path back to the entrance. There seemed to be an inherent reversibility built into the structure of a unicursal maze.

So how to make sure that a visitor would be trapped irreversibly?

As the evening set in, his tiredness returned. He fell asleep at his desk, eventually, and slipped into a remarkable dream.

He was walking inside what he knew to be a labyrinth. The path he was treading looked like an ordinary narrow alley, but he had an unexplained feeling (as one often does in dreams) that it was not a regular alley. It was the interior of the labyrinth. He could see daylight above, and he was squeezed between two narrow walls, much taller than he was. Suddenly he became aware that his pockets were rather heavy. When he checked, he found out that they contained ten round coins—shiny metal coins with a circular maze depicted on both faces. They had a sinister look. Perplexed, he kept walking onwards. He was taken by great anxiety as he proceeded—a sense of claustrophobia oppressing his mind.

After a short while, he reached a door that blocked the way forward. Its robust wooden surface looked impenetrable. There was no keyhole, no handle: nothing other than plain, homogeneous wood. He started searching the whole surface meticulously, seeking some crack or other small opening that might allow him to get through. It was only when he looked up at the sky, in desperation, that he noted a thin aperture to the right of the door, in the stone wall. A flash in his mind showed him the coins in his pocket. The aperture seemed tailor-made for them. Feeling suddenly more hopeful, he reached for a coin and inserted it into the slot. The door opened at once, onto what looked like a continuation of the path he had been following. He walked through. The door closed immediately behind him. As he was walking hesitantly away, he heard a tingling sound: the coin he had placed into the slot on the other side had fallen through and was rolling towards him on the ground. It was no longer shiny. It had turned as black as coal. Instinctively, Ts'ui Pên scooped it up and went on.

Nine times this situation repeated itself, and nine times Ts'ui Pên inserted a shiny coin into the slot by the door, walked through, and was given a dark, tarnished coin back.

Shortly after the tenth door, he reached a circular courtyard: but sadly, a dead end. Seeing that there was nowhere else to go, Ts'ui Pên hurried back onto the path, toying with the coins in his pocket, aiming to reach the tenth door, which he had so recently passed through. He still had the chance of retracing his steps and returning to where he started because there had been no bifurcations along the way. After a little while, the door he had gone through at last reappeared. Confident of his previous experience, he reached for the slot on the left-hand side, to insert the coin. Again, as often happens in dreams, he felt he already knew what to do. He inserted the coin, but nothing happened. The door remained firmly shut.

The anxiety came back, even more powerful than earlier. He felt like he was freezing in the deepest parts of his gut. He was trapped with no hope of escaping, his heart racing as he fumbled with the slot, trying to insert the coin over and over again. Finally he fell to the ground in despair, and he lost his senses.

He awoke at his desk, the light of the early-morning sun shining brightly through the gaps around the heavy curtains. Minuscule speckles of dust were scattering the light all around, moving about in their unstoppable dance. Despite the horrid feeling of being trapped still hovering in his mind, and his heart still racing, he was secretly pleased. The dream had showed him how to answer the riddle. It is, in fact, easy to create irreversibility in a reversible, unicursal labyrinth. It is enough to require some resources to be necessary to go through doors that otherwise block the visitor. If the visitor is enticed to enter the labyrinth and to continue towards the centre, hoping to find an exit at the far end, it will be easy to encourage him to consume the resources, given enough doors, and cause him to be trapped. In the dream, it was the shiny coins that opened the doors, not the tarnished ones. By depleting the shiny coins, Ts'ui Pên had been trapped with no hope of escape, despite the path having no bifurcations.

This quirky experience is, allegedly, what sparked Ts'ui Pên's deep interest in mazes, and led to his subsequent obsession, described in "The Garden of Forking Paths".

This is more or less the Earl of Kendal's account of the lost story about Ts'ui Pên. As I said, I am not sure of its veracity; but, after all, is there anything that we know for sure? I had many other conversations with the Earl after that one, but we never came back to that topic. Then one day he disappeared, without saying a word. No one in University College knew anything about where he had gone. He had simply vanished without leaving any trace or clue as to where he might have gone. I long wondered what had happened to him, until one snowy morning I found an envelope in my pigeonhole, and in the envelope, a letter. It was handwritten in crimson ink, and it was accompanied by a rare edition of a book on labyrinths. The letter contained a brief, kind note signed with his name. It said he had gone on a journey, to visit all the remaining ancient labyrinths left in Europe and the Middle East, searching for a unicursal labyrinth that is not reversible.

He has not come back since.

7.

A Journey There and Back Again

> ... *me al largo*
> *sospinge ancora il non domato spirito.**
>
> —Umberto Saba, "Ulisse"

At the beginning of this book I made an ambitious claim: I promised to present a new perspective on physical reality, by explaining how to understand, and think in terms of, counterfactual properties of physical systems. It is now time to head back to where the journey began, and to contemplate the new understanding that we have gathered along the way.

---

*From "Ulysses": "... *far offshore / I am still driven by the untamed spirit.*"

With the last page approaching, one has to operate like a thoughtful traveller nearing the end of their journey: review the things that seemed noteworthy; consider what new avenues they open up and whether they are fuel for the creation of further knowledge.

The central, motivating idea of this book is that a class of properties has been largely neglected in science; and that this needs to be remedied because it is preventing progress on fundamental problems.

I called those properties 'counterfactuals': they are not specifiable by describing the actual state of a physical system nor its law of motion; to specify them, one has to describe the system in terms of what transformations are ultimately possible, or impossible, to perform on it. As I have explained, counterfactuals are lurking at the core of most existing open problems in fundamental physics. I mentioned several examples: the interoperability of information (chapters 3 and 4); the no-cloning property and the reversibility of quantum information (chapter 4); the resilience of knowledge (chapter 5); the conservation of energy and the distinction between work-like and heat-like energy transfers (chapter 6); the information-based interoperability of work media in thermodynamics (chapter 6). The main counterfactuals you have explored in this book are summarised in the table that follows.

Fundamental Counterfactuals and
Their Relation to the Physical World

Phenomenon	Counterfactuals	Physical Laws	Notable related entities*
Information	*Possibility* of 'flip' and 'copy'	Interoperability of information media	Universal computer Universal constructor
Quantum information	*Impossibility* of copying certain information-carrying states; *possibility* of reversing any transformation	Interoperability of information media	Universal quantum computer Universal constructor
Knowledge ('resilient information')	*Ability* to enable transformations and of remaining embodied in physical systems	To be discovered!	Abstract catalysts Universal constructor
Work	*Possibility* of the 'seesawing' transformation *Impossibility* of changing energy without side effects (*conservation of energy*)	Conservation of energy Interoperability of work media Counterfactual second law of thermodynamics	Scale-independent (micro and macro) heat engines Universal constructor

Having now an overall perspective, one might notice that two general facts emerge—the two overarching ways in which the power of counterfactuals expresses itself.

---

*For these entities to be *possible*, one needs the counterfactuals in the second column to be allowed by the laws of physics.

The first is that adopting counterfactuals brings entities that look superficially like immaterial abstractions into the domain of physics. Information and knowledge, for example, have been traditionally considered as mere abstractions—as things that do not belong to the physical world. However, by considering the counterfactual properties of physical systems that enable information and knowledge, one refutes this idea: because whether or not a physical system has those properties is set precisely by the laws of physics.

The other fact is that embracing counterfactuals allows one to express exact laws about entities traditionally considered as approximate (because these laws refer directly to the macroscopic world), such as information, energy, heat, and work. When the counterfactual properties enabling those entities are made explicit, it is elegant and easy to express laws about the systems displaying those properties, without approximations.

This is how the power of counterfactuals allows one to ground concepts that would otherwise be considered abstract, or approximate, in exact, fundamental physical laws.

The logic of how that is done is the same for all the entities discussed in this book—it is a unifying trait. First, one expresses the counterfactual property that is required of a physical system for it to embody the entity in question. For instance, if the entity is information, the counterfactual properties that the system must possess are the possibility of the flip and the copy transformations. Then one can express regularities about the physical systems with those counterfactual properties in the form of laws of physics—for instance, interoperability laws.

For example, a 'bit' of information looks like a pure abstraction until we view it through counterfactuals. Then one notices that for a system (e.g., a switch) to qualify as a bit of information, it must have two counterfactual properties: one, that it is possible for it to be in either of the two physical states ON and OFF; and, two, that the state ON and the state OFF can each be permuted into one another and also copied into any other physical system that itself has these same two counterfactual properties. Whenever a system has these two counterfactual properties, it qualifies as an information medium.

You see, then, why a bit is not an abstraction, independent of the physical world: whether or not something has those properties depends entirely on the laws of physics. Unlike, say, whether a given number is a prime number, which does not depend in the least on what the laws of physics are. Those counterfactuals provide the link between information and physical laws.

In addition, one can state an interoperability law about the system that is displaying those counterfactual properties. The interoperability law explains why the counterfactual properties, though physical, are not dependent on most of the details of the physical systems: it is because, when there is an interoperability law, those properties are shared by a class of physical systems. The details of the systems all belonging to the same class become irrelevant.

In the case of information, the interoperability law says that by putting together two systems each qualifying as a bit, you will get something that itself behaves like an information medium, with the same counterfactual properties as each bit (permutability and copiability) being present in the composite system. This is true irrespective of how the bits are embodied: whether it is in photons, electron spins, or binary switches. That is why information can be copied from any such system to any other.

Whenever there is an interoperability law, some details of the physical system obeying the law are not relevant to the entity being expressed and can be abstracted away. In the case of a bit, many slight variants of the states of the physical system carry the same information—for instance, an arrow pointing straight up or down, compared with an arrow that is very slightly tilted to the left or to the right. They all have the same property: that this information (and not necessarily the variations in the state) can be copied into other physical systems. All other details of those physical systems are irrelevant for the copiability property. For instance, a switch with the states ON and OFF is a bit, but so is an arrow with the two orientations UP or DOWN. The irrelevant physical details, such as how worn the switch is, or how thick the arrow is, or its colour, do not affect the counterfactual property in question, and so one can leave aside the details of how the bit is realised. But enabling those counterfactual properties depends on the laws of physics, so they are physical properties after all, despite being abstractions.

As I have explained time and again, the traditional conception of physics cannot express counterfactual properties. The traditional conception can refer to the state of the switch—either ON or OFF—at any given time, and can predict what the state will be at a later time, and why. However, a statement of this kind does not tell us anything about what transformations are possible or impossible on the switch. This is why turning to counterfactuals and related laws is essential to capture the physics of phenomena such as information and the other entities you have discovered in this book.

There is another unifying aspect of the approach I have been advocating, which was foreshadowed at the end of chapter 2. All the counterfactual properties you have encountered are expressible as statements about which transformations are possible and which are not, and why. A daring speculation is therefore that all the laws of physics could be formulated solely in terms of principles about counterfactuals, and that the laws of motion follow from them as derivative, and perhaps approximate, properties. Exploring this possibility is the start of an exciting research programme. To develop it, one would have to formulate laws of physics about systems displaying the counterfactual properties discussed in this book, and to show that dynamical laws such as quantum theory and general relativity are emergent, derivative approximations following from those principles. This is potentially a whole new avenue for physics, being opened up by taking counterfactuals seriously. It is for physicists—and other scientists and philosophers—to explore its developments in the years to come.

A programme of that sort is no simple matter. For a start, in order to adopt any of these putative laws about counterfactuals as laws of physics, one would have to ensure that they are *testable*.

A law is testable if it produces predictions about observable traits of a physical system. Testability, as I mentioned in chapter 2, is a pillar of the scientific method, where a theory that makes testable predictions can be refuted if its predictions are not borne out by experiments. A classic example (as recounted in chapter 2) is testing a prediction about the speed of a ball of a given mass rolling down a slope with a given inclination. Economics and medicine are two disciplines where testability is problematic because, although predictions can be made, it is hard to make repeated experiments under controlled conditions to check the predictions against reality. Physics, on the other hand, is privileged because many of its predictions are testable.

Can principles about counterfactuals be tested? Yes, but the process is one step removed from the tests of rolling-ball predictions that you saw in chapter 2. Principles such as the conservation of energy are in general tested by deducing their implications for the behaviour of physical systems that are assumed to obey them. Principles are *laws about laws*—they are meta-laws. One needs first to have at least two rival theories concerning a physical situation to which the principle purports to apply. For instance, one can consider a model for a pendulum that obeys the principle of energy conservation—for example, a model based on Newton's laws—and then another model that does not, predicting that once the pendulum is set into motion, it spontaneously swings to higher and higher points. Then one performs an experiment with an actual pendulum, to test the prediction of one model against the other. In the case of the pendulum, all experiments done so far have refuted the model that predicts that energy is spontaneously created. Whenever it looks as though the pendulum swings to higher and higher points, it is because some nearby system is actually providing the energy to do so—for instance, by driving the oscillations with some mechanical engine, which provides the require energy for the swings. So far, the principle of conservation of energy has withstood all tests performed on it in its domain of applicability.

In the history of science, there are well-known cases in which a principle was tested and there were significant consequences. The discovery of the neutrino is an example of the power of general principles: the process of decay of a neutron into a proton and an electron seemed to violate the principle of conservation of momentum. In fact, what was discovered is that a third particle with almost zero mass is produced, the neutrino; this was postulated by Wolfgang Pauli and later verified by experiments.

Principles about counterfactuals are tested in precisely that way. For instance: the principle of interoperability of information requires that if we put together two systems that each behave like an information medium, then the composite system should also be one. This statement predicts what types of transformations must be possible for the composite system, once we know that certain transformations are possible on each one of its component systems. One way to test this principle is to consider two rival models for, say, two bits embodied in a particular type of physical system. One theory predicts that the principle is violated because once they are put together, they cannot be used to store two bits of information. And it explains why—because some of the copy-like operations from one bit to the other are not allowed—just like in the example I gave in chapter 3, with one bit being made of 'dust' and the other being made of ordinary matter. The confirmation of one or the other model would therefore refute or corroborate the interoperability law for information. This is why principles about counterfactuals can indeed be testable, thus satisfying the crucial requirement of testability for laws about them.

The principle of interoperability also has an intriguing twist, which allows one to make predictions about physical systems without knowing exactly what their laws of motion are. It is something I have been working on for the past few years. With dynamical laws, the only way to make predictions about what happens when two systems are considered together is to know the dynamics of each of them, as well as how to construct the composite laws of motion. But sometimes one does not know all that—for example, it is still a matter of heated debate whether and how gravity obeys quantum theory; the reason is that the current best explanation for gravity is general relativity, a theory with no quantum information media in it. So we may not have all the tools to describe the joint motion of a qubit interacting with gravity. Yet we may still want to make certain testable predictions about that system. Other times, even if the law of motion is known, it is too complicated to follow all the motions of the constituents of one of the systems—this is the case for complex molecules that have so many subparticles that it is impossible to use the laws of motion to predict their behaviour: too complicated even for the existing supercomputers. In such cases, counterfactuals come in very handy because they can allow us still to make predictions. Why? Because they hold for those systems irrespective of their details! Imagine you know that two systems each qualify as a bit. But you do not know all the details of their dynamics, nor how to describe the dynamics when they interact with one another. The interoperability of information would still allow you to make predictions about certain tasks on the composite system—because it is based on the counterfactuals rather than on the dynamics. This is an example of how the laws about counterfactual properties can be useful and go beyond the testable predictions of known dynamical laws.

I want to mention a very recent example where this logic applies, which is at the heart of the current struggle to merge the two best explanations of the universe known to us, quantum theory and general relativity. There are some physical systems, such as particles with masses comparable to those of human cells, for which both gravity and quantum theory are thought to be relevant. Yet there is no unified dynamical law that describes a system that both gravitates and is quantum. There have been brilliant proposals to achieve that unification, but to date, none of these candidates has been conclusively chosen over the others. When it comes to those systems, we do not know what law of motion we should be using to make predictions. Still, we know that the counterfactual interoperability law applies in that domain, even when the specific laws of motion are unknown. So we can use the interoperability law to make predictions in that domain! This approach with counterfactuals has recently led to an idea to test effects in quantum gravity, which has created a lot of interest within the quantum gravity and the experimental communities. The race to realise the experiment has started; and if it is realised, it could finally refute the idea that gravity is not quantum. Such is the reach of counterfactuals: they provide a powerful underpinning of deep conceptual ideas, as well as the robust theoretical structure to support experimental ideas of this kind. I expect that there will be more experimental ideas to come.

In addition to the exciting consequences for physics, switching to counterfactuals has deep, important implications for other fields. One of them could be to revolutionise the understanding of knowledge.

I said that knowledge is a particular type of information, with the counterfactual property of being resilient—it can cause itself to remain instantiated in physical systems. I also explained that we do not know exactly how it is created, but we know that it can arise out of no-knowledge via the process of natural selection, and that another process for creating new knowledge is what happens in the brain when we think.

Science does not know if there are new laws that govern this type of resilient information. But rooting knowledge in counterfactuals is the right approach to creating a corpus of such laws. This is not least because the approach via counterfactuals frees knowledge from all the subjective connotations that have traditionally plagued theories of it.

Moreover, with counterfactuals, knowledge becomes a physical entity—rooted in the resilience of a particular kind of information, which is an objective counterfactual property independent of observers, sentient beings, and the like. The most far-reaching consequence of this shift is that some open problems that have been traditionally labelled as spiritual, mystical, and even religious— such as finding laws and regularities about knowledge and its evolution—can, via that shift, be posed firmly within the scientific domain, without appeal to dogmas or to supernatural ideas. This is the first necessary step in order to solve those problems via scientific methods, and it relies on counterfactuals.

This switch removes from the shoulders of scientists and rationalists a heavy burden that comes with those problems—an apparent dilemma, which goes like this: On the one hand, there are certain phenomena that require explanation—phenomena such as artificial selection, the unfolding of creativity at the level of the individual (with new theoretical ideas popping up in various disciplines as a result of individual creativity) and of society (with the progress of civilisation). We are intrigued by these phenomena and compelled to understand them in depth. Yet contrary to this intuition, it is often anathema to scientists to talk about creativity in human brains, knowledge, and related phenomena as having any real significance. This is because of a prejudice that affects the sciences in much the same way that prejudices of other kinds affect religious thinking. Knowledge is regarded suspiciously as anthropocentric, subjective, and related to Descartes's 'mind-body' dualism, which is the root of all sorts of misconceptions that are also deeply entrenched in religious thinking. As a result, many open problems about the human mind and knowledge creation are sometimes regarded as not interesting by some scientists. The contemplation of the possibility of laws applying to knowledge and the like appears like literal nonsense to part of the scientific community: some retreat to the domain of reductionism and materialism, denying that knowledge is a phenomenon requiring an explanation. Others simply ignore the question, thinking that it goes several steps too far into stuff that is not proper science.

But counterfactuals provide a way out of this trouble. If we use counterfactuals to define knowledge as information with the ability to last (such as that in genes that code for useful adaptations), and creativity as the ability to create new knowledge, we are able to free them of any subjective tinge, making them objective. That is still very far from providing a theory of those phenomena, but it provides a scientific handle on knowledge by grounding it in the laws of physics.

After taking this step, one can make further useful moves. First, consider as a problem for physics the fact that certain systems in the biosphere exhibit a property that no other system in the known universe has. Namely, creativity—the ability to create new knowledge by thinking. Human brains have this ability. (It may be that other systems—say, beetroot plants—have it, too, but it is, at best, much less manifest than it is for humans.) Now, thanks to our objective definition of knowledge, stating facts like this— that human brains can create knowledge—is no longer vulnerable to allegations of anthropocentrism. Claiming that it is anthropocentric would be equivalent to considering the statement "A dishwasher has special properties among all the known systems in the known universe" as supporting a dishwasher-centric view of the universe. Clearly, it does not. It is an objective statement about the fact that dishwashers can do certain things (such as quickly scrubbing away dirt from cutlery and crockery and being unchanged by the experience) to an accuracy that is unparalleled in the known universe. Likewise, the statement "Human brains are capable of constructing new knowledge" is not anthropocentric.

The relevant difference between the case of human brains and that of dishwashing machines is that the detailed functionality of being knowledge-creating is not as well understood as the functionality of dishwashing machines. Still, via the counterfactual notion of knowledge, one can now refer to knowledge creation in an objective fashion. It is the first essential step in regarding problems about creativity as pertaining to the domain of science, which is in turn step zero to even hoping to address them effectively.

Another consequence of this switch is that we can consider a number of other related issues as coming into the domain of science. There are several examples. First, there is the issue of whether there are other systems in the universe with the same creative capabilities as the human brain: other species on Earth; other forms of life on other planets; or, possibly, existing and future artificial intelligences. Related to this is the problem of understanding how knowledge comes into the world—understanding the thinking process and creativity from the point of view of physics and information theory. Starting from our objective definition of knowledge, we can wonder properly within science about questions such as: How can one set up that creative ability in a computer? Is the termination of creativity with death necessary, as some would like to argue, or is it actually possible to defer death to later and later in life, by means of some sort of error correction? What is it, exactly, that we have to store and error-correct? Can a person be copied, stored, and downloaded into another embodiment when the time of death for the body approaches?

Each of these topics requires very careful consideration, and it could be the subject of countless research programmes. Here I just want to point out one fact. Whatever your view on these questions, it is far easier to approach them from the point of view of science, free of the prejudice that knowledge-related issues are anthropocentric, subjective, and nonscientific. Counterfactuals allow one to take all these issues seriously and fully within the scientific method.

From a more parochial point of view—the point of view of the survival of our species—these questions are vital for a simple reason. Among all the things we do, some are short-lived, such as eating, sleeping, brushing our teeth, and taking care of the various materials that our body is made of, whereas others have the ability to last, because they have to do with creating entities that, like information, can be moved from one physical embodiment to another without changing their salient traits. Creativity is one of the main tools we have to form stuff that can last. If one is interested in making the good outcomes of our civilisation last and improve, then understanding how creativity is nurtured—both at the individual and at the societal level—is essential. As the poet Mary Oliver put it:

> There is within each of us . . . a third self,
> occasional in some of us, tyrant in others. This
> self is out of love with the ordinary; it is out of
> love with time. It has a hunger for eternity.

It is crucial to take seriously that third self, and nurture it in everyone; and for that purpose, it is essential to know what creativity is—whether it is a thing at all, or just an illusion as some would like to say. Counterfactuals, and the idea of knowledge as resilient information, offer science a way to take creativity seriously and to aspire to a theory of it rooted in physics.

I am now imagining you approaching the end of this book. You are, to my eyes, much like an explorer returning home. The harbour lights are finally in sight. After having traversed previously unexplored waters, you are coming to the end of this journey.

The new perspective of counterfactuals can present you with a last gift before you close this book. It may modify your very perception of endings.

Let me start with the ending of an adventurous expedition—a good analogy with the ending of a book. To tell the truth, such endings have often had a bad reputation. It is a trope that there is a sense of anticipation and excitement before a quest; a sense of tension while the quest is ongoing; then, once everything is over, even if the quest has been successful, there may be a melancholic sense of loss.

Epitomising this pattern is a little-known poem, "Alexandros", that speculates about the thoughts of Alexander the Great as he approached the Arabian Sea, at the end of his dazzling military campaign to conquer Asia Minor. The author of the poem is one of the most accomplished voices of nineteenth-century Italian literature, Giovanni Pascoli.

Alexander was the king of Macedon in antiquity, in the fourth century BCE. He is one of the most influential figures in ancient history—perhaps in the whole history of humanity.

Starting from his homeland, Macedon, Alexander completed his father King Philip's conquest of Greece, and then part of the Balkans, and Egypt, and the entire Middle East, finally arriving at the Indian Ocean. He managed to perform this task in a little less than ten years, mostly in his twenties. Before that, he had been taught by the great ancient philosopher Aristotle in a private school founded by his father King Philip.

The ancient historians created an aura of magnificence and mystery around Alexander. They also tell the tragic story of his death. He died abruptly at the age of thirty-one, possibly poisoned—though the exact cause of death is unknown. His intense and brief life stands as a testimony to how remarkably productive one can be when devoting oneself single-mindedly to a sole objective.

It is Pascoli's speculation that, contrary to what one might expect, at the end of his journey Alexander is profoundly dissatisfied. In the poem, Pascoli deploys several tools to evoke the haunting memories and visions in Alexander's mind as he reflects on the path he has just trodden. Alexander has gone through fantastic adventures, learnt fascinating facts, and won many battles. Yet he is discontented, according to the poet's interpretation, because he thinks that there is nothing more to discover. "Dreams"—

the poet comments in a couple of lines—are the "infinite shadow of reality".

Put more plainly, this lapidary comment implies that Alexander would have done better to stay at home, just dreaming about how his future military campaigns might unfold, rather than engaging and going through the journey to conquer the world in reality. The reason is simple, according to Pascoli: while dreams can continue ad infinitum, because they are never confronted with reality, the military campaign has to unfold and come to an end. Even when the campaign is successful, its end means that there is nothing more to do in that regard; there is nothing left to look forward to.

But is this really true of all endings? Not necessarily. Whether or not an end is positive or negative depends on its counterfactual properties—which brings us back to the main theme of this book. Endings can be positive points for further construction, provided that knowledge has been created along the way—so that the end is richer in possibilities (in counterfactuals) than the start. Once more, the literary tradition comes to our aid with an example. It is the account of a journey where the beginning and the end, for the character undergoing the journey, coincide. A journey 'there and back again'. This type of circular journey has a special name in literary jargon: *nostos*, from an Ancient Greek word (νόστος) meaning 'return'. When the *nostos* is successful, the character reaches a satisfactory new state upon completing the journey. In this case, the end of the journey is the richest and most promising moment in the story.

One of the most famous *nostoi* is that of Odysseus. It is also one of the first *nostoi* to have ever been told—according to the tradition, by the Greek poet Homer (just like Euclid, Homer may or may not have existed as an individual; but the tradition treats him as a real person, and so shall I).

Odysseus, an ancient king of Ithaca—a Greek island in the Ionian Sea—left home to join the Greek armies fighting against the city-state of Troy during the Trojan War. The war lasted about ten years, during which those armies kept Troy under ferocious siege.

His is the mind that crafted the notorious 'Trojan Horse'—a hollow wooden horse to be left in front of the gates of Troy as an offering, pretending that the Attic army had given up on the siege. The horse's hollow body would be hiding a handful of the bravest Attic soldiers. The hope was that the Trojans, having seen that the Attic army had vanished, would open the doors and bring the horse inside the city walls, as a trophy. And so it happened. Once within the city walls, the Attic soldiers hidden inside, led by Odysseus, crept out unnoticed at night and opened the gates to the rest of the Attic army, who razed Troy to the ground.

After the victory, the Achaean princes set out for home. These princes were not exactly saintly, and they all had several bad deeds to their names, which is why (according to the myth) some of them were punished severely by the gods. A few managed to get home quickly, but others were brutally killed along the way. Odysseus's *nostos* was unique: successful in the end, yet particularly long and laboured; blessed all along by the creation of knowledge.

Homer's account of Odysseus's *nostos*, the *Odyssey*, is a haunting epic. The lasting fame of Odysseus is largely due to his counter-factual abilities, which have to do with creating knowledge. During the journey, he is tested at different levels. Only if he is sufficiently smart and strong of mind will he be able to reach home without losing himself. Unlike Alexander, as portrayed in Pascoli's poem, Odysseus has not lost himself at the end of the *nostos*, but has acquired more knowledge and has the ability to put it to some future use. The end of the *Odyssey* has a firm, higher point than its start, which is essential for further improvement—it is richer in possibilities, which are special cases of counterfactuals. Alexander, by contrast, has lost his capacity to dream of further deeds; his creativity—and all the knowledge acquired along the way—is therefore useless, because it cannot be put to any use.

More generally, any journey, without necessarily being literally an exact circle, can still be a successful *nostos* of sorts, and its end can be a positive, uplifting fact. What matters is whether, along the journey, the character has or has not managed to create more knowledge while preserving his or her own individual capacity to create new knowledge. So an ending can be a fertile starting point; it depends on whether the character reaching the end is still capable of being creative. In fact, a successful *nostos* does not have an ending. Its ending is the starting point of new adventures.

Just as with a *nostos*, there may be no actual ending for a book, either, provided that something particular occurs while the reader goes through the book. When a reader makes his way through a book, a unique relationship is established between the book and the reader, which grows within the space that the writer Philip Pullman has masterfully called the 'borderland'. The relationship is entirely based on counterfactuals—the knowledge that the reader creates in their mind while reading the book; and it is something unique and private to the reader. Whenever knowledge creation happens while the book is read, the reader undergoes a *nostos*: even once the book is completed, implications of the knowledge created along the way stay with the reader for as long as the reader's mind survives and can be put to some use in the future. In that case, the book does not really have an end.

As I am writing these lines, I am thinking of you, my reader, like a modern Odysseus. You are now emerging from this journey, approaching the end; you can moor the boat at the wharf, take a well-deserved rest at the hostel that overlooks the harbour, and look back through your memories, considering them carefully.

As you are pondering all the ideas encountered in the book, perhaps a smile lights up your face. Before you, there are still vast, unexplored waters waiting for you to take to the sea once more and create further knowledge.

May the knowledge you discovered in this book serve you well on the journey.

Alexandros

The gatekeeper was resting on his chair just outside the temple. He was guarding the temple because the great king Philip of Macedon had recently opened a school there. It was no ordinary school: it was open only to a few boys from noble families. The king's own son, Alexander, attended it.

The only teacher at the school was Aristotle. The legendary philosopher had moved a few years earlier all the way from Athens to Mieza, just to fulfil the request of the king.

At that time of the day, the villagers stayed well hidden in their houses, keeping out of the heat. It was the perfect time to rest. The gatekeeper was contemplating the prospect of a few hours of quiet, when he saw a small figure rushing up the temple steps.

The figure was dressed all in white—it was a boy wearing the classic tunic of the school's pupils. The gatekeeper immediately recognised him as the son of the king.

As soon as he saw who it was, the gatekeeper relaxed back to his state of quiet. Alexander's visits to the school after hours were nothing new. In fact, Alexander seemed to have some sort of arrangement with Aristotle to talk to him outside the scheduled classes. The gatekeeper was not sure how that ten-year-old boy could be interested in even more schooling, but he imagined it was some sort of supplementary class arranged for him because he had to train specifically to be the future king. Little did he know that Alexander had reached a far better deal with Aristotle. Those meetings were not classes in the traditional sense. The purpose was to have conversations *with the great philosopher. The boy did not enjoy the regular classes that much; but the conversations were a different matter. They were open-ended, exciting, and far-reaching.*

This is how, that day, Alexander came to be rushing up the marble stairs, as quickly as his ten-year-old legs allowed, with his heart beating rapidly as he was running uphill—even faster than the exertion alone would have caused, because he was thinking that he would soon be talking to Aristotle.

As the gatekeeper settled down to his siesta, Alexander slipped into the temple.

To Alexander, the interior of the temple meant safety from chaos and disorder; it meant quiet; it meant time for his mind to become engaged and enthused. Inside, everything was still and silent.

Aristotle was waiting for him at his usual place. He was sitting in a small garden, bounded by ordered pillars, in the shade of scented lemon trees and colourful oleanders. A large white dog was lying fast asleep on the floor, by his chair, breathing quietly and regularly.

"You cannot know how delighted I am to see you again, dear boy", said Aristotle emphatically. "Without you, it is just me and the dog, who, though friendly, cannot sustain any dialogue. Come here, sit down and tell me everything."

Alexander wetted his lips with the tip of his tongue. He always felt a rush of anxiety before exposing his thoughts to Aristotle. But he managed to say, all at once: "My question today is about what it is that makes me enjoy all this poetry and philosophy and maths and abstract stuff that you are teaching us. I figured if I'm made of the same stuff as rocks and grass, why do I need to have an understanding of these other, abstract things, that don't have any particular embodiment?"

Aristotle gave him a long, deep look. Quite a question from a ten-year-old! The boy had certainly engaged his attention. The philosopher was now sitting straight in his chair, suddenly as alert as a lizard in the presence of prey.

"Well, I do not know!" he exclaimed. "We don't understand the world yet, my dear boy, let alone the mind. We're far, far from understanding an infinitesimal part of it. We will always be, I suspect." *He chuckled, smiling briefly to himself. The idea that the stuff yet to be known could be an infinity no matter how much one knows amused him.*

Then he continued: "What's clear is that the mind has characteristic properties that make it capable of relating to things that are abstract. I suspect that it obeys the same laws as rocks and grass, though we have yet to find these laws and understand how to apply them to the mind."

Alexander did not say anything at first, because he could not yet think of anything useful to say in response. His mind was filled with fluttering thoughts, each following a possible implication of that provocative statement. Then finally he managed to articulate one of them—the one that bothered him most:

"But what is it that has such a strong hold on me when we discuss abstract things? Why can I focus on things that seem not to be real—things made just of words and reasoning and airy thoughts, not of gold or silver or other material substance?"

"You think deep, Alexander, son of Philip. You will go far in life", said Aristotle sententiously. Then he fell silent for a while, looking away, searching for the right words. That boy was asking hard questions. "The stuff you enjoy, Alexander, is what we call knowledge. It is something abstract—but abstract does not mean 'unreal'. It is part of reality as much as that pillar over there and this dog over here. As you can see, we can move knowledge from one mind to another, and also create it, or sometimes even destroy it. But it is intangible, which is why we say it's an abstraction.

This is why it is hard to notice it, and to get to like it. To do so, you have to cultivate a special love for thinking."

Alexander was mesmerised. "Create? How do we create knowledge?"

"Do you ever find yourself daydreaming? Fantasising, making up stories in your mind? Or perhaps with your mind completely absorbed by an issue, a problem, so that you forget about everything else around you? That is when you are most purely thinking. When you think, you create new ways of looking at the world around you, to solve problems—new explanations. What I try to do in this school is to inspire that critical attitude towards reality that makes you think and solve problems as fast as you can. That is essential to the process of creating new knowledge, which I once learned from my master, and he from his."

"Indeed, I am critical", Alexander thought swiftly. "I'm asking you lots of questions, and you have no clue how to answer!" But he held his tongue, politely, and just kept listening.

"You see", Aristotle continued, "in the future, there will be times when your choices will make the difference between the life and death of people under your rule; between the prosperity of your kingdom, and its collapse; between you yourself being happy or unhappy; and so on. My mission is to help you be ready for when those days come. I am, in this respect, your loyal servant."

Alexander smiled widely in response to that little oration. He liked the idea of being in charge. He felt important and responsible when Aristotle said words of that kind. He felt that he wanted to give himself wholeheartedly to the cause of being the best king of all times. He added mindfully: "I wonder what is the best training to gain the right knowledge? How do I make sure I can create the best knowledge possible?"

Aristotle shook his head slowly. "Well, unfortunately, it is not possible to be absolutely sure of anything; let alone to know for sure what the 'right' knowledge is. We cannot have the guarantee that anything we know is true, or that some piece of knowledge is better than another. It's interesting you are asking these questions. I have only recently realised these deep facts; actually, they refute some of the ideas that I have already published in my writings. But I will remedy that: I will explain to the world, I hope, how things are. I have to get round to doing it. It takes time." Aristotle started patting the dog, which was sleeping at his feet. The dog opened one eye, wagged its tail a little, then went back to nap. Aristotle took no notice but kept patting the dog and talking. "The most important thing is that you learn to have a critical attitude and an open mind: if you make a mistake, which you will, you can rapidly try to correct it. You have to learn how to be unsatisfied with stagnation; to be constantly looking for problems to solve; to question whatever rule or idea is imposed on you, trying to find a valid criticism of it, to make it better. You can always try to improve on things by conjecturing tentatively better solutions. But everything is tentative, not absolute—that's the key to making progress."

Alexander liked that idea. It was nice to hear that it is all right to make mistakes and that the goal is to realise as quickly as possible that one has made a mistake. Still, he was not sure how to connect that with the notion of knowledge. He tried a daring statement to see if he had guessed correctly: "I see. So you say, Aristotle, that my best bet for being able to do all this critical thinking is to train my mind to like these abstractions—to like knowledge and the activity of creating it?"

"Yes. I'd say, my dear boy, that every evil is due to a lack of knowledge. So if you want to be prepared against evil, try to be prepared to create knowledge and improve on things tentatively, by making mistakes and trying to correct them as fast as you can. Be open to your own mistakes and ready to try to correct them promptly."

"You gave me a nice idea", the boy said suddenly, his face all lit up and cheerful.

"What is it?"

"I would like to build a gigantic house in some fastness, protected from wars and evils, where everyone can come and read things so that they can learn how to enjoy knowledge—a fortress for books. I want books to be saved from the madness of evil men. I want knowledge to survive across very many generations so that everyone can access it, and criticise it, and try to find mistakes in what has been written down by the previous generations. Wouldn't that be a wonderful thing?"

"What a luminous thought, my dear boy! It is very wise and very studious of you. You wish to fund a house for books, to protect them. This is what I believe should be called a library. You could name the library after yourself—the Alexandria."

The boy smiled widely again and jumped slightly and clapped his hands with a triumphant expression. "It's a perfect name. Actually, I have an even better idea. I can use that name for a city. I shall command its construction and that of a great library within it."

Aristotle smiled softly. "He-he. That may take some time, Alexander, and you may be carried away by other duties. But I hope you will keep this love for knowledge. And that you, or someone close to you, will build a library like that. As I said already, lack of knowledge is what causes every evil on this earth."

"Is it because if we knew things, we would not repeat our mistakes?"

"Yes. Think about it. If we knew better, we would know how to cure diseases, how to improve the quality of life, how to improve our morality. Humanity could be better off in so many ways."

"But what about death, then? That is certainly due to natural causes, not to lack of knowledge. What can we do about it, after all?"

"What now appear to be inevitable natural causes may be averted by the knowledge that has yet to be created. You see, you'd be burned by the sun if it were not for this tunic you are wearing, and for the walls of this temple, which are constructed to provide

shelter to us against atmospheric agents—the sun, the lightning, and so forth. The tunic and the temple are made out of our knowledge of how to get the fabric out of materials that come from the animals in our flocks; and of how to manipulate rocks into becoming materials for architecture. The sun may be inevitable, but its harmful effects on us need not be. I look ahead in the future—perhaps centuries away—and think that we can make intellectual progress as fast as we can now. I could imagine that what now looks like a very challenging disease to cure will become easier and easier to deal with. Strange as it may seem, even retarding death may be only a question of understanding things better. The same holds for the mind. One day we will know how it works, and why in certain respects it is just like a rock, and in others utterly different; and how it can create knowledge and at the same time obey the same laws as all other objects that are physical. We will be able to answer your questions fully; or to know why it is impossible to do so if there are some limitations to that that we currently don't know about. And we shall have a host of new and even better questions."

Alexander was silent for a while, contemplating what Aristotle had just said. It seemed fascinating and counterintuitive; hard to embrace, but hard to contradict. So, in the end, he opted for a side comment: "I wish I were able to see this progress myself."

"We don't have a choice of when to live. We only have the choice of what to do with the time we have. Choose wisely, Alexander, son of Philip."

Alexander could listen forever to this type of talk. He looked ahead, focusing on an indefinite point, lost in contemplation of this far, faraway future. His gaze was suddenly conscious and wise.

"Well, Aristotle", he said with a serious expression on his face, finally looking the philosopher right in the eyes, "I will try to make my story worth telling, and to make the most out of my time."

"And so you shall", Aristotle said in a whisper.

He seemed tired of the chat: Alexander knew when it was time for the meeting to come to an end. He swiftly said: "I will go for now. I will think of more questions and come back soon."

Aristotle looked at him with a gentle, benevolent gaze: "Good. You can always visit me, especially if you have questions of this kind, which for me are the greatest of delights. As I said, between you and me: I am planning to rewrite completely my works about poetry, morality, and aesthetics, in the light of these considerations about knowledge. I hope there will be time for me to do that. I will do it soon. These discussions are helping me greatly to order my thoughts."

Alexander was flustered by the praise. "Very well, I will make sure to have more interesting questions."

"Ah, one more thing", said Aristotle. "I wish to give you this." *With those words, he handed Alexander a large scroll that he had produced mysteriously from his tunic.*

"What is it?" the boy asked with surprise, holding the codex in his hands as if it were a sacred token from the gods.

A Journey There and Back Again

"*It is a copy of the* Iliad *by Homer—my annotated copy.*" *Aristotle smiled. "It is one of the finest pieces of poetry about knowledge, war, and love, and all those things that a future king like you may want to know. Take it, Alexander, it's yours.*"

A few moments later, as the gatekeeper was again looking out at the hills on the horizon, now flecked with the hues of the sunset, he noticed the boy—the Prince, the son of Philip. This time he was trotting down the temple steps. In the light of the evening sun, the boy's shadow looked long and imposing, like that of a great and noble man.

Acknowledgements

I am very lucky because several ideas presented in this book were refined through extended conversations with brilliant friends and collaborators: David Deutsch, with whom I have had a wonderful collaboration about several topics at the foundation of physics—specifically on constructor theory and related ideas; Vlatko Vedral, who has been a source of the most sceptical and most supportive feedback, and has perceptively illustrated this book; and then Peter Atkins, Simon Benjamin, Harvey Brown, Keith Burnett, Paul Davies, Geri Della Rocca de Candal, Artur Ekert, Angelina Frank, Karl Gerth, Francesca Loria, Philip Pullman, Paul Raymond-Robichaud, Anicet Tibau Vidal, Maria Violaris, Sara Imari Walker, and Albert Wenger. I particularly wish to thank Albert, David, Peter, and Vlatko for providing sharp criticism on earlier drafts of the book. I am also grateful to Rowan Cope, Laura Stickney, and Paul Slovak for improving this book in many essential ways with their insightful suggestions; and to John and Max Brockman, who have been supportive of my book project since its infancy.

Acknowledgements

More broadly, I thank the exceptional teachers and lecturers who sparked my love for physics, literature, maths, and philosophy, turning my education into a fantastic adventure: Nella Cogno, Luciana Armellini, Adele Scattina, Ersilia Castelnuovo, Anna Maria Dell'Anna, Maria Paoletti, Luisella Caire, Paolo Tilli, and Giovanni Monegato.

I also wish to thank my mentors—David Deutsch, Artur Ekert, and Mario Rasetti—for the knowledge that we have discovered together; and for championing the idea (which this book intends to celebrate) that the pursuit of scientific discovery relies on a combination of freedom, responsibility, and intellectual delight.

I thank my mother, Piera, for providing constant encouragement in regard to this book project, and for having sparked my interest in languages (and in several other beautiful things) when I was a child.

To Vlatko goes my thanks for accompanying me on this journey; for our endless conversations; and for making this Multiverse a much more interesting and loving place.

Essential Further Readings

Atkins, Peter. *The Four Laws That Drive the Universe*. Oxford University Press, 2007.

_____. *Galileo's Finger: The Ten Great Ideas of Science*. Oxford University Press, 2004.

Brockman, John, ed. *Possible Minds: Twenty-Five Ways of Looking at AI*. Penguin Press, 2019.

Dawkins, Richard. *The Extended Phenotype: The Long Reach of the Gene*. Oxford University Press, 1982.

_____. *The Selfish Gene*. Oxford University Press, 1967.

Deutsch, David. *The Beginning of Infinity: Explanations That Transform the World*. Penguin Books, 2011.

_____. *The Fabric of Reality*. Penguin Books, 1997.

Pearl, Judea, and Dana Mackenzie. *The Book of Why: The New Science of Cause and Effect*. Penguin Books, 2018.

Pullman, Philip. *Dæmon Voices: Essays on Storytelling*. David Fickling Books, 2017.

Vedral, Vlatko. *Decoding Reality: The Universe as Quantum Information*. Oxford University Press, 2010.

In Memory of My Father

I think that my father would have enjoyed reading this book. He would have certainly challenged me about some of the more controversial ideas, gently and sharply, as he used to. When talking to him during our seemingly never-ending walks (we used to walk, and walk, and converse about almost everything), I forged a few of the themes of this book. Between the lines, you can find his relentless enthusiasm for science; his passion for knowledge; his love for storytelling. Almost every object in my childhood home had a curious story—he was a master of imagination: Where does that come from, why is it there, and where is it going? These marvellous fantasies made my childhood dazzling with enchantment. I hope you can find a little of that in the short stories herein. There is also his irony, of course; his playful attitudes towards life. He never took himself too seriously; but took seriously all those things above; and the love for our family, and the knowledge within our family traditions. Turning these things into prose is my own way of remembering. In the current impossibility of having him back, this seems to be the next best possible thing.

Index

Index

computation, ix, 28–29, 31; copying of, 84; elementary, 91–92; examples of, 89–92; and laws of physics, 89–94, 109–10; quantum, 35, 125, 128; universality of, x, 92–93

computer science, xix, 81, 89

computers, 70, 82; bits of, 110; and carrying information, 79; classical, 109–10, 173; copy operation and, 84, 120; and counterfactual properties, xvi, xviii, 76, 88–89; creative ability in, 219; and heat engines, 195; interoperability property and, 90–91, 96; and laws of physics, 2, 89–94; memory of, x, 84, 86; programming of, xvi, xviii, 91–93; quantum type of, 35, 78, 89, 93, 125, 134, 173, 194; repertoire of, 90–92, 94; unfairly regarded, 88–89. *See also* universal computer; universal quantum computer

constraints, 2, 33–34, 134–38, 167, 195

contradictions, 50, 117–18, 176

Copernican Revolution, 20

Copernicus, Nicolaus, 20

copiability property, 33–34

copy operation, 209, 213; abstract catalysts and, 151–52, 156; and carrying information, 76, 82–86, 96; centrality to computers, 84, 120; description of, 76, 84; DNA molecules and, 10–11, 84, 146, 148, 150; far-reaching power of, 103; impossibility of, 105, 125, 136–38, 206; information media and, 96–98, 111, 119–20; and manuscripts, 3, 11, 101; and measuring anything physical, 120–22, 130; from one support to another, 11, 88; possibility of, 34, 76, 82–85, 96–97, 129, 206–8; Sentosa bridge, 82–85; story about, 99–104. *See also* information: copiability; printing; replicators

cosmology, 48, 55–56, 94

counterfactuals: adoption of, 207–11; barriers to, 25–28; description of, ix–xii, xv–xx, 4, 31–35, 205–7; far-reaching power of, 98, 128, 215–17; fundamental ones, xvi–xix, 206; how to approach, xix–xx; importance of, xvi, 31, 34–35, 42, 207, 210; links between, 96, 120–21; new research on, 210–11; and positive endings,

223–25; reformulating physics with, 68, 174; unfairly regarded, xviii–xix; and unifying traits, 207–10. *See also* *specific properties*

counterintuitive facts, 118–19, 126, 139–42, 190, 235

creativity, 7, 16, 18, 29, 41, 51, 217–21, 225

crystals, 97, 114–19, 123

Cumean Sibyl, legend of, 44–45

Darwin, Charles, 9–11

Dawkins, Richard, 10

death, 173, 219, 222, 234–35

Descartes, René, 217

design, appearance of, xix, 34, 154–55

'deterministic nightmare', 62–63

Deutsch, David, ix–xii, 68, 89, 93

DNA, 5–6, 8–11, 78, 84, 146–51

'Dust sector' (hypothetical), 93–95, 213

dynamical laws, 168; and counterfactuals, 59, 61–62, 66, 210; explanations by, 31, 42, 54–55, 64; and external entity of time, 57–59; and making predictions, 63, 66, 213–15; new research on, 210–11; and principle of impossibility, 61–62; problem with, 27, 59, 61–63, 67, 73; reversibility of, 52–53, 59, 183; story about, 70–71

Einstein, Albert, xviii, 16, 22, 71, 73, 108, 128

Ekert, Artur, 118

electromagnetism, 53, 77

electron spins, 87, 129, 209

electrons, 1–2, 32, 40, 77, 212; attraction by electrostatic force, 143; in computers, 28; and conservation of energy, 168; and entanglement, 126; and laws of motion, 67; as qubits, 123; and seesawing transformation, 192; velocity/position of, 121–22

electrostatic force, 143

emergent entities, 31, 34, 88

End of Time, The (Barbour), 58

Ende, Michael, 133

endings, perceptions of, 221–25

energy, 22, 79, 119, 143; definition of, 167–68; exact laws about, 207; heat-like transfer of, 30, 171–73, 175–78, 180–83, 186, 193, 195, 205; interoperability of, 170–71; irreversible transfer of, 171–74, 177, 179–83, 193; and relation to mass,

Index

Index

Heisenberg uncertainty principle, 122, 127
Higgs particle, x–xi
His Dark Materials (Pullman), 94
Hofstadter, Douglas, xii
Homer, 224–25, 237
human brain, 34, 82; electric currents/
  signals in, 2, 77, 169; information
  copied into, 84, 87; knowledge and,
  13, 163; particles of, 65; thinking
  process and, 216–19; universal
  computers and, 93
human mind, 14–17, 20, 49, 155, 174, 217,
  226, 229–33, 235

IBM, 110, 125
Iliad (Homer), 237
impermanence, 1–4
impossibility: and conservation of energy, 61,
  165, 167, 170; of copying, 105, 125,
  136–38, 206; and counterfactuals,
  61–62, 65–67; laws/principles about,
  61–62, 68; of predictions, 115; and
  quantum physics, 122, 127, 129; story
  about, 198–203. *See also*
  transformations: impossibility of
industrial revolution, 30, 172
infinite regress, 55–59, 67–68, 167
information, ix, xii, xx; and connection to
  physics, 76, 78–86, 88, 95–97, 208;
  copiability, 86–88, 96, 146, 148–50,
  153, 206, 209; and counterfactuals,
  xii, xix, 11, 34–35, 76, 87–88, 97–98,
  206–8; in DNA, 150–51; exact laws
  about, 207; freed from subjectivity,
  97; fundamental unit of, 85–88, 91–92,
  94, 110–13; how to express exactly,
  173; instantiated, 155; interoperability
  of, 93–96, 205–6, 209, 212–14; local
  inaccessibility, 136–37; non-
  copiability, 94, 206; in physical
  systems, 13, 77–86, 189, 194–95; and
  possibility of 'flip', 206; reductionists
  and, 29; resilient, 13, 216; self-
  preservation of, 149, 151; special
  cases of, 11; story about, 99–104;
  technology, 195–96; and traditional
  conception of physics, 42, 207
information media: classical, 128–30;
  copiability, 119–20; counterfactual
  properties of, 85–89, 96–97, 105; and
  general relativity theory, 128;
  interoperability and, 94–95, 98, 189,

206, 209, 213; need for two
  counterfactuals, 208; non-quantum,
  109, 128; quantum, 111–13, 122–23,
  128–30, 214; reversibility and,
  122–25; work media as, 194
information processing, 109–11
information theory, 68, 112, 119–21, 129–30,
  195, 219
initial conditions, 34, 38, 107–8; cannonball
  example, 52–55; of computers,
  28–29; counterfactuals and, 67–68;
  description of, 25–32, 52; dynamical
  laws and, 31, 58, 62–64, 67, 73; and
  game of chess, 64–65, 68; infinite
  regress and, 55, 58; and laws of
  motion, 25–30, 32, 52–58, 60–62; as
  open problem in physics, 56–60
interactions, naturally occurring, 143–44
interoperability property: computers/
  computations and, 90–91, 93–96;
  description of, 87–88, 213–15; and
  energy, 170–71; and information
  media, 87–88, 98, 205–6, 208–9,
  212–14; and making predictions, 215;
  for mechanical means, 188–89; testing
  of, 212–14; and work media, 206
irreversibility: at core of physical reality,
  173; counterfactual, 165, 185–88;
  'forgetful', 183–84, 187; and second
  law of thermodynamics, 177–79,
  183–86; statistical, 165, 183, 187;
  three different kinds of, 165; and
  transfer of energy, 193; and
  transformations, 186–87; of
  unicursal maze, 199, 202
IT companies, 110

Joule, James, 185, 193

Kelvin, Lord, 186
Kendal, Earl of, 197–203
knowledge: and abstract catalysts, 139,
  152–53, 155, 157; abstract nature of,
  230–31, 233; Aristotle on, 230–37;
  and catalysts, 139, 142–45, 149; and
  counterfactuals, xii, xx, 206–7;
  definition via counterfactuals, 1,
  34–35, 155–57, 216–21; description
  of, 13, 43, 144–45; and laws of
  physics, 155–57, 218–19; mole
  cricket's den story and, 139–42,
  144–46; resilience of, 13–14, 17, 163,

Index

particles: elementary, x–xi, 1–3, 8–10, 26, 28,
53, 66, 88–89, 123–24, 143;
interactions of, 21, 64–65, 143, 179;
with masses, 215; microscopic, 29,
40, 174, 192; minuscule, 168;
quantum, 118, 127; subatomic, 66, 77
Pascoli, Giovanni, 221–23, 225
Pauli, Wolfgang, 212
perpetual motion machine, xvi–xvii, 2,
61–62, 66
photons, 20, 87, 209; and counterfactual
properties, 119–20, 122–25, 129; and
crystal experiment, 114–19, 123; and
entanglement, 126; information-
theoretic properties of, 119–20; in
quantum states, 112–18; as qubits,
123; and reversibility, 123–24; and
seesawing transformation, 192;
splitting of, 114–15; in superposition
of different paths, 114–19, 121–22
physical reality, 20, 23, 173–74; and
counterfactual properties, 34; narrow
take on, 28, 63; new perspective on,
204; and physical systems, 78;
theories that explain, 26, 52, 77;
transformations occurring in, 149–50;
understanding of, xv, 24, 29, 32, 96
physical systems: carrying information,
77–86, 189, 194–95; and catalysts,
148–49; composite, 125–26, 209,
213–14; and conservation of energy,
167–70; constraints and, 134–38, 167;
counterfactuals of, xvi–xviii, 33–35,
78–88, 166, 204, 207–8; factual
properties of, 79, 85; information in,
13, 155; of information media, 85–86;
macroscopic properties of, 184; as
made of bits, 85–86; and music boxes,
165–71; not carrying information,
78–79, 85–86; properties needed for,
81–85; states of, 79, 81–83, 85, 181–84,
193; transformations on, 151
physics: approximations in, 22–24, 30–31,
187–88, 193, 207; classical, 113,
128–29; and counterfactuals, xvi, xx,
65–66; and formulating explanations,
18–21; fundamental, x–xi, xv, xviii–xix,
21, 26–28, 30–34, 53, 178–80; open
problems in, xx, 27–28, 56, 67, 205,
216–17; point of, 19–21, 36–41
physics, laws of: and counterfactuals,
xvi–xvii, 1, 4, 8, 27, 67, 209–10;

description of, 1–3, 38–41; equations
and formulae of, 22–23; essential
features of, xvi–xvii, 46, 143;
fundamental laws, 178–80; new
research on, 210–11, 215–16; next
revolution in, 41; preserving
elementary components/
interactions, 154; second law's status
in, 177–80. *See also* no-design laws
physics, traditional conception of, 60, 167,
185, 207; boundaries set by, xix;
counterfactuals not expressed by,
32, 42, 97, 210; counterfactuals
departing from, ix–xii; and
dynamical laws, 42, 56–59, 63; and
knowledge theories, 216
Planck, Max, 186
planets, 4, 42–43, 66; and Newton's laws, 16,
46–47, 54; orbits of, 2, 16, 144; other
forms of life on other planets, 219;
and solar system, 19–20
Poincaré, Henri, 182–83
Popper, Karl, 14, 16, 155–56
possibility, xvii, 22, 34, 68, 79. *See also* copy
operation: possibility of;
transformations: possibility of
predictions, 23, 119, 121; conjecture/criticism
and, 25–26; and counterfactuals, 44,
59–60, 214; dependence on
underlying explanation, 45–47, 49;
description of, 43–47; and
dynamical-law approach, 63, 66;
examples of, 43–47, 53, 65;
experiments about, 47, 49–52,
211–12; and Galileo's motion theory,
49–51; impossibility of, 115; and laws
of motion, 53–54, 213–15; and
Newton's laws, 46–47; and second
law of thermodynamics, 194; strict
criteria for, 47–49; testability of, 25,
47–49, 51–52, 56, 67, 211–14; and
traditional conception of physics, 32.
See also unpredictability property
primitive elements, 20–21, 35
principle-based approach, xii, 66–67,
71, 73
principles, description of, 211–12
printing: of books, 99–104; history of, 197;
machines for, 101–4, 196; of
newspapers, 84
problems: and counterfactuals, 54, 68;
holding promise of improvement,

Index

Index